# THE SCIENCE & ART OF MANAGEMENT & EXECUTION OF INDUSTRIAL PROJECTS IN INDIA AND OTHER DEVELOPING COUNTRIES

Anand Kumar Gupta
and Luiz Baptista

Copyright © 2019 Anand Kumar Gupta & Luiz Baptista

All rights reserved.

*To our wives Arunima Gupta and Elsa Baptista, who assumed total responsibility of all home affairs, by and large single handedly, through long years, thus enabling us to successfully handle our careers as project managers; it being, more than a full-time job so far as attention of a Project Manager and his/her responsibility and answerability is concerned.*

# ACKNOWLEDGEMENTS

## By Anand Kumar Gupta

I am most indebted to Dr. N.C. Saha, Professor of Electrical Engineering at College of Engineering & Technology and Dean of Faculty of Engineering, AMU Aligarh, who inculcated in me the habit of trying to understand matters of engineering as problems of physics and mathematics and trying to find their solution through application of physics and mathematics.

In course of my career as an engineer I got properly oriented towards 'Engineering Practice' under valuable guidance of Mr. G. C. Jain, General Manager of Renusagar Power Project of Hindalco Industries Limited. It was during the course of my working in this project where I could get out of my inhibition of being electrical engineer and not a mechanical engineer or a civil engineer. I feel deeply indebted to Mr. Jain for opening my eyes to the fact that in real life all branches of engineering converge into physics and mathematics and therefore with open mind and intelligence all engineers can handle all problems successfully.

I am indebted to all colleagues in Torrent Power Limited who supported me in all my whims and fancies and worked hand in hand with me, during execution on EPC basis by a reputed international company, to handle frequent changes in execution schedules depending upon circumstances almost on a day-to-day basis to ultimately make both the combined cycle power projects a success.

Finally, I am deeply indebted to corporate management of Torrent Power Limited for listening to my reasoning and logic in selection of technology for its two combined cycle power projects and also giving me a free hand to shape tender technical specification and EPC Execution contracts. This resulted in the two power projects being remunerative as intended.

# ACKNOWLEDGEMENTS

## By Luiz Baptista

I would like to show my profound appreciation to my dear friend Mr. Gupta who challenged me on this very pleasant endeavour to co-authorship this manuscript, filling me with satisfaction in being able to share my professional life experiences with younger generations also involved in this fascinating art of managing engineering and construction projects.

Finally, I am very grateful to my family who wholehearted supported me during the long period of preparation of this contribution. The many days and nights of "being absent" from the normal routine of family life and constant contact with my dear ones were often frustrations they endured with love and compassion, knowing how important this task was for me.

# CONTENTS

1. Preface     1
2. Introduction     4
3. The Human Aspect - The Project Manager's Fitness     14
4. Project stakeholders and their role in success     27
5. Project Contracting     34
5.1 Nature of Contract and Number of Contractors Responsible for Project     36
5.2 Clarity about Deliverables and Scope related to Deliverables     45
5.3 Legal Responsibility of Project Owner and/or Contractor/(s) towards Laborers     57
6. Project Execution     65
6.1 Handling Manpower (Lessons from India)     71
6.2 Sourcing goods & machinery from developing countries     78
6.3 Execution of actions for meeting contractual requirements     92
6.4 Taking over project site to start project construction work     94
6.5 Civil construction work - Preliminary and preparatory work     97
6.6 Starting physical work at construction site     110

| | |
|---|---|
| 6.7 Management of Civil Work | 122 |
| 6,8 Management of Steel Structural Work | 135 |
| 6.9 Procuring of Steel Sections | 140 |
| 6.10 Management of other erection activities inside structural building | 146 |
| 6.11 Management of erection of tanks, vessels and pipes | 148 |
| 6.12 Management of Installation of various Equipment, Plant & Machinery | 160 |
| 6.13 Special features of installation of Control & Relay Panels and Cables | 162 |
| 6.14 Installation of cable trays and cable carrying conduits | 164 |
| 6.15 Installation of electrical equipment, plant and machinery | 167 |
| 6.16 Management of installation of Mechanical Equipment, Plant and Machinery | 169 |
| 6.17 Management of Commissioning of Industrial Project | 174 |
| 6.18 Activities post Commissioning and completion of Performance Test | 184 |
| 6.19 Management of Handing Over of project to owner. | 185 |
| 6.20 Spare parts & tools | 188 |
| 6.21 Post-Performance Test handling of facility | 189 |
| 6.22 Responsibilities/Services Post Handing Over of Project | 190 |
| 6.23 Warrantees during warranty period | 191 |
| 6.24 Long Term Service Agreement (LTSA) | 193 |
| 6.25 Closing of project execution contract | 195 |
| Appendix 1: Culture Effect | 196 |
| Appendix 2: Types & tendencies | 203 |

| | |
|---|---|
| Appendix 3: Managing Health and Safety Performance | 208 |
| Appendix 4: Managing Project Risks | 213 |
| Appendix 5: Ensuring Quality | 217 |
| Appendix 6: Effective Project Control | 222 |
| About the Authors | 226 |

# PREFACE

In June 2011 I accepted the invitation, to be Construction Manager, of a reputed European Original Equipment Manufacturer (OEM), in a large Engineering Procurement Construction (EPC) power plant project in India, already in construction. At that time, I thought that due to my previous and successful assignments with similar responsibilities, it would be just another challenging project in another unfamiliar country like many others I had experienced until then. During my mission there, which ended in February 2013, I realized that in fact I was living an amazing professional and spiritual journey which lit me inside in several aspects related to my professional activities.

The aim of Mr. A. K. Gupta (in short AKG) through this precious manuscript was originally to disclose a collection of practical hints related with the science and art of Project Management in India, which Mr. Gupta, then my client's Senior Project Manager, understood might be relevant to project manager readers to get acquainted to. I was kindly requested by him to review this manuscript also putting my previous long experience in this matter, taking advantage of my native background from another developing country (Brazil). I promptly accepted and started immediately this pleasant journey along this manuscript. I realized though that most of the testimonies Mr. Gupta put in light were pretty much similar to those I had experienced

1

in similar industrial projects held in other developing countries, namely in South America and Asia. I shared this observation with Mr. Gupta who immediately extended his invitation to me in order to become co-author in this lovely adventure, which I accepted, with the mission to add my own accumulated lessons learned along my professional life as project / site manager elsewhere, which due to similar local challenges makes these lessons also applicable in both arenas. Therefore, the reader will find the views and testimonies from two experienced project managers in their respective fields of expertise and cultural background, who decided to extend their sincere friendship, no matter the impediment of the geographic distance which separates our lives, towards the common goal of sharing their own lessons learned along their professional lives, with others who may be in similar profession.

I think it is worth to mention to the readers also my admiration of Mr. Gupta's superior character and expertise in the art and science of project management, as well as his profound love to his incredible country. His virtues acknowledged by all his peers too, made my decision to co-work this manuscript an honor and at the same time gave me a huge pleasure. I was lucky enough to have had him as my mentor along my memorable assignment in India, nurturing me into the secrets of Indian way of thinking and values. Despite the fact that our respective companies were sometimes sitting on different sides of the table, on issues related with that particular project execution, later on I happened to learn from him that most of these disputes could have been avoided or substantially mitigated if the alleged "unpredictability" or contingencies of working in India, and by extension in developing countries too, would have been considered on the early phases of the project by means of better understanding of local practices and cultures, most of them addressed in this manuscript.

The text covers all project phases, since the contracting till the closing, and is rich of examples, best practices and awareness of pitfalls present along the project execution, some of them regrettably learned too late. At the appendices, the reader can also find relevant general management topics that global project managers also have to take into consideration to avoid unpleasant surprises along the project execution. Of course, 100% predictability is not a realistic approach anywhere, due to the unique dynamic nature of occurrences in projects and its circumstances. In realistic scenarios, when project managers are confronted with high level decisions, one should never forget to consider also the holistic and strategic approaches when managing projects, which goes beyond the pure technical perspective only by means of budgets, progress control and quality, which are the most familiar to project managers. To be also considered are the social (people) and institutional (society) perspectives, directly related with social environment where the project is executed, along with other key components of project, challenging project managers to be vigilant in these matters as well.

As final words, this manuscript is a collection of hints and best practices, which configures a big support for those project managers acting in India and with ambition to act in the global arena, mainly in other developing countries, and reported by two experienced and passionate professionals, who believe that managing projects effectively is not only making use of classical technical tools and procedures, but much beyond than that, by means of science to understand the truth and of art to apply the science.

Wish you to enjoy reading it as we did writing it.

- Luiz Baptista (LB)

# INTRODUCTION

The first thing that I owe to reader of this work is to provide a small explanation about why I chose the words 'science' and 'art' in the title. During my days as an Electrical Engineering student at Aligarh Muslim University (years 1959 to 1963) one of the books that we were recommended to refer to in connection with Protective Relays was titled "The Art and Science of Protective Relaying", authored by C. Russell Mason of G.E. USA. I still remember that Mr. Mason had written in its preface somewhat to the effect as below.

> *Science is knowledge of truth and Art is application of Science for benefit of humanity.*

Ever since becoming an engineer, my endeavor has been to first look at things scientifically and to understand them scientifically and then to use my instincts and put the best of art bestowed on me by God or nature for ensuring that whatever I do is ultimately beneficial to those who may be concerned with it. In title of this work I have purposely used the word "science" first followed by the word "art", reason being that unless the person concerned has first scientifically understood any subject or matter or thing, he/she would most probably not be able to put his/her art to work towards beneficial use of that subject or matter or thing.

Having said that I would also like to admit that there is no dearth of very good books on the subject of Project Management, all authored by acclaimed experts capable of teaching me the subject of Project Management. Therefore, with all humbleness I wish to go on records that I do not consider myself to be an expert of Project Management, in the conventional sense as I do not possess any specific qualification on the subject of project management. Therefore, it has been my endeavor to steer clear of reinventing the wheel. It often so happens that despite the best available wheels fitted to best available vehicle, the vehicle is not able to successfully traverse through the voyage. Obviously, there is need to talk about the science and art of successfully completing a voyage, given that best wheels are already fitted in the best vehicle.

My purpose of authoring this work is to just put forth my interpretation of what I observed during my journey in life entailing power projects, more so with reference to execution of various types of Industrial Projects in India since the state of art as of date does involve human physical as well as intellectual input to a very large extent – as project construction, erection and commissioning of equipment with project handling capability, in totality – including humanitarian aspects, is not available. If this work is found useful by any one, my efforts would be justified.

I am an Engineer with experience of 50 years in Operation and Maintenance of Power Plants and in managing Power Plant Projects, starting from the year 1963. First three years of my experience as an engineer were related with testing, commissioning and subsequent maintenance of Protective Relaying and Power Line Carrier Current (PLCC) communication systems. Subsequently, I got chance to work, on owner's side, in a coal based captive thermal power plant project comprising two units, being executed on turn-key basis, in Uttar Pradesh, India, by a highly reputed company of USA. Considering the state of art prevailing at that time the power plant being constructed by the company from USA was of the most advanced technology, till then not seen in India. The project comprised of two generating units and work at site had started in mid-1965.

The first generating unit was commissioned and put in to successful commercial operation in September 1967 followed by the second in early 1968. One gets experience from facing situations posing serious risk of failure and succeeding to come out of them without fatal damage. It is like succeeding to float back after getting in a situation of drowning. Since project work is always one of a kind and each project is unique by way of being executed at a site that is different, the project team is different and desired end results differ from other projects. Each project is full of surprises and chances of failures with slightest error of judgment but success each time lies in ability to quickly surmount the surprise before it becomes a story of failure.

I was a learner at that time and very keenly observed the methodology of management and execution being followed by the company from USA. Subsequently; I was involved in Operation and Maintenance (O&M) of this power plant and ultimately headed its O&M setup. It so happened that this power plant faced several failures and root cause of these failures was found to be discrepancy between calorific value and ash content of the coal that became available for firing vis-à-vis the figures of calorific value and ash content of the coal based on which the power plant was designed and engineered. As a result, coal crushers, coal pulverizers, equipment for firing the coal in boiler furnace, boiler heating surfaces and flue gas paths, ash collecting system, ash handling equipment and system were all found to be inadequately designed and engineered.

It took 2 to 3 years to gradually improve designs of all these systems and equipment one after the other and only after about 4 years from commissioning the power plant achieved high reliability and performance so much so that its availability and annual average output factors reached levels unbeaten by any other power plant in India at that time. Thereafter, the plant required only regular maintenance and routine operation, leaving negligible scope for innovation and the work in a sense had become rather repetitive and boring.

Luckily in 1977 I got a chance of working with a Consulting Engineering Company as Project Coordinator for a 3x110 MW + 2x220 MW capacity coal based thermal power project in Rajasthan. However, in 1979, I was approached by the previous company to join it as Chief Engineer responsible for engineering and execution of expansion project of the captive thermal power plant that was executed by G.E. USA, with the purpose of adding two more generating units and three pulverized coal fired boilers so as to have five boilers to feed four turbines.

The first two units were designed on the philosophy of Unit System so that each turbine had a dedicated boiler. However, since the boilers were not of the reheat type so that each turbine was of once-through type rather than being reheat type. The OEM had provided a steam and feed-water pipes cross-over system such that the main steam as well as feed water were provided with interconnecting pipes with valves such that in case boiler of say, Unit-1, has a tube failure and at the same time the condenser of turbine of Unit-2 has a leak, rather than both units having to be shut down the boiler of Unit-2 and turbine of Unit-1 could be operated together in Unit System thus preventing total loss of power generation. This was a very clever and thoughtful engineering. I considered the project of addition of two turbo-generators and three boilers as a golden opportunity for me to conceptualize the project, improvise it and carry out necessary engineering, in-house, and to manage the carrying out, in-house, of the necessary procurement, construction, erection, testing and commissioning and putting into successful commercial operation of the project.

The work of project engineering was started in mid-1979 and completed in 1980 and construction started in 1981 to be completed in 1983-84 with successful putting of the project into commercial operation. My innovation was in conceptualizing parallel operation of five boilers to feed four turbines as a result of which load on each boiler would be only 80% of its designed capacity while each of four turbines would operate at 100% of its designed rating. Since the rate of erosion of components due to fly-ash in flue gases is in exponential proportion to flow velocities of gases in boiler, each boiler having to run only at 80% of designed rating meant drastic reduction of wear and erosion of boiler components and also much better performance of the Electrostatic precipitators.

I am glad to mention that this project not only got executed successfully well in time and well within the estimated cost but also resulted in totally non-stop 100% power supply by four generating units resulting in exceptional profits to the owners.

This was a boon in disguise as any lack of sufficiency in terms of correctness of engineering, quality of construction material and power plant machinery and compromises made during execution of the projects would have been the responsibility of my set-up to answer. It gives me great satisfaction to mention here that such exigencies didn't arise, giving me confidence about the systems and methodologies adopted by my set-up for execution of the projects. It resulted in the parent company deciding to create its own setup somewhat like EPC, of which I was given the charge to handle. The internal EPC setup so created later successfully handled several power projects pertaining to expansion of business.

During my tenure described above, I learnt about Murphy's Laws. Murphy's Law has become a generic term for any statement that is witty and cynical. For benefit of the reader are quoted below the 14 variations of Murphy's Laws that I have found most applicable to industrial projects.

- If there is a possibility of several things going wrong, the one that will cause the most damage will be the one to go wrong. It is also expressed as, "(i) If anything can go wrong - it will or (ii) Whatever can go wrong - will go wrong".
- If there is a worse time for something to go wrong, it will happen then.
- If you perceive that there are four possible ways in which a procedure can go wrong, and circumvent these, then a fifth way, unprepared for, will promptly develop.
- If everything seems to be going well, you have obviously overlooked something.
- Nature always sides with the hidden flaw.
- Whenever you set out to do something, something else must be done first.
- Every solution breeds new problems.
- Enough research will tend to support your theory.
- When there is a very long road upon which there is a one-way bridge placed at random, and there are only two cars on that road, it follows that: (1) the two cars are going in opposite directions, and (2) they will always meet at the bridge.
- Smile... tomorrow will be worse.
- Left to themselves, things tend to go from bad to worse.
- You never run out of things that can go wrong.
- All's well that ends.
- A meeting is an event at which the minutes are kept and the hours are lost.

Experience during the above cited tenure has taught to me that whatever be the reason the Murphy's Laws do have some reason for being proven true from time to time and the science behind it is Theory of Probability and all can be mathematically derived.

I owe a lot to my timely getting real life experience with the Murphy's Laws, for subsequently being able to successfully execute projects during my earlier mentioned tenure and gradually it has almost become my habit to first try and mentally analyze the situation in light of Murphy's Laws.

Subsequent to my tenure up to the year 2002, however, I got a chance to work for a company in connection with its two large Combined Cycle Power Plant based on natural gas. Here my role was limited to (i) selection of technology, (ii) framing technical part of specifications for international competitive bidding for execution of the project on EPC basis, (iii) framing technical part of EPC contract. A multinational company of very high global repute, also having its establishment in India, won EPC the contract.

My role, thereafter, was to take care of owner's interest at respective project sites to ensure that after being put into commercial operation the respective plants would operate reliably, trouble free and with very high availability – all at commercially viable costs. Thus, my position for these two projects was that of owner's project manager.

In view of there being a multinational company as EPC contractor and recollecting my early experience with G.E. as well as Siemens Germany, as a young engineer, initially I was under the impression that all would go very smoothly with these projects and execution will be eventless and well in time. However, both these projects suffered perhaps all that can go wrong. For sake of making it clear as what did go wrong, I am enumerating below most mentionable of things that had gone wrong.

- Safety related incidents,
- Poor quality of work – despite all good intentions of the EPC Contractor,
- Quality issues in supplies (mainly bought-out items),

- mostly since personnel concerned of the EPC Contractor took matters on face value,
- Bankruptcy of subcontractor, death of owner of subcontractor firm leading to key personnel switching over to other firms and cash crunch with the subcontractor,
- Undue delays in completion of civil works to the extent of making it ready for placement and erection of equipment,
- Early dispatch of equipment from manufacturers works demanding additional Lay-down Area for uncalled for storage at site,
- Deterioration in quality of equipment due to unduly long site storage owing to early deliveries compounded by delays in completion of civil works,
- Misinterpretation of provisions of technical specification by the EPC contractor and corresponding repeated discussions to reach to correct interpretations,
- Transit damages and damages during storage and handling at site,
- Undue delays in commissioning due to undetected failure of sensors leading to multiple useless visits of experts from abroad – all to no avail until someone got a flash in mind to change the sensors,
- Unduly high consumption of gas turbine fuel (free issue by owner in terms of the EPC contract in the first of the two projects) due to lack of presence of mind on part of the commissioning personnel of EPC contractor – leading to totally avoidable cost overruns on owner's side without any fault on part of the owner,
- Extra expenditure on part of EPC contractor due to the above and corresponding demand of EPC contractor for owner to share such extra costs even though there had been no default on part of owner,
- Poor evaluation of sub-contractors, driven almost exclusively by initial prices rather than long-term values,
- Lack of proper identification by EPC contractors of key

cultural local diversities, as well as factoring the element of unpredictability in some cases.

I warned the persons concerned at EPC contractor's end based upon my observations at site and work schedule and analysis of these in light of Murphy's Laws. However, my warnings and advices were ignored as the EPC contractor's persons being a part of the OEM did not expect the cultural differences between attitude of Indian persons and that of other countries and did not expect that there would be numerous matters including cultural in India which they had no knowledge about. Worst was that although the persons of the EPC contractor's team did feel that something was wrong they still expected that things will happen in normal course as they happen all over the world and most probably hardly any one tried to analyze the situation on grounds of a cultural difference and attitude of Indian people. This added to probability of errors and omissions.

Having said as above I wish to admit that whatever I have mentioned in this work is more of a recollection of my experience and almost none at all pertaining to the theoretical topics related with project management such as PERT (Program Evaluation and Review Technique) and CPM (Critical Path Method) on which there already exist plethora of books authored by eminent and highly accomplished experts with tons of qualification and knowledge on their back.

I also do not intend to cover accounting and budget keeping aspects since there is hardly any worthwhile commercial organisation that has not already implemented good enough computerized system of accounting and expense control. Therefore, trying to invent the wheel will rather be counterproductive.

My purpose of authoring this work is to be able to help the grass root project managers, mainly those dealing with international projects and contractors in India, who may have faced failures despite their excellent educational background and availability of lot of theoretical material on Project Management.

Feather in the cap is that my friend Mr. Luiz Baptista kindly accepted my request to be a coauthor of this book. His kind acceptance has enhanced reach of the book so that its contents have become useful to project managers even outside India.

In the end, I would like to say that since my real-life experience of executing projects has been limited to India, I have titled my work as such but I am sure that Indians are of the same human species as the people anywhere else in the world. Therefore, no lesser than 95%, of whatever I have tried to put in this work as the extract of my lifetime experience, will be equally beneficially applicable to any part of the world.

- Anand Kumar Gupta (AKG)

# THE HUMAN ASPECT

*The Project Manager's fitness*

What is a Project? How we react when faced with the responsibility of sailing an industrial project to success?

Oxford Dictionary defines Project as: An individual or collaborative enterprise that is carefully planned to achieve a particular aim.

Cambridge Dictionary definition is almost identical: A piece of planned work or an activity that is finished over a period of time and intended to achieve a particular aim.

Wikipedia, the dynamic encyclopedia and current favourite of many, defines project elaborating a little further as: Planned set of interrelated tasks to be executed over a *fixed period and within certain cost and other limitations*.

It is of special interest here to take a note of the underlined portion of Wikipedia definition of project. Three specifics are seen in Wikipedia definition viz., (i) the expression "fixed" that qualifies 'period', the expressions (ii) "within certain cost" and (iii) "other limitations" too are of extreme importance for understanding the definition of 'project'. Unless, the above said three qualifiers, (i) 'fixed' for "period", (ii) 'within certain' for "cost" and (iii) 'with other limitations' as an overall qualifier, are fully understood and strictly adhered to a project of any cannot be executed successfully i.e., to serve the intended purpose ... fully and truthfully.

So far as the person who is responsible for execution of a project is concerned, the above three expressions "fixed period", "within certain cost" and "notwithstanding other limitations" (the word 'notwithstanding' purposely added for clearly elaborating the boundaries that bind all projects and cannot be crossed).

While on the subject, it would be worthwhile to look at what all of us perform in our day-to-day lives. Let us take a look of some very familiar engagements or events of our day-to-day lives:
- Cooking food for family,
- Dropping kids to school and picking them back at school,
- Traveling with family for vacations,
- Celebration of birthdays and anniversary ... etc.

There are certain other more complex performances, which we are all either familiar or able to somewhat imagine how they take place:
- A stage performance, particularly a Play,
- Carrying out a complex surgery by a team of surgeons and other medical experts.

A little analysis of each of the above, depending on which ones of the above you have personal experience of handling, will make you recall the advance planning, efforts, finance management, management of odds, recovery from failures, events which gave pleasure and satisfaction of success or misery of failure. Whatever, be the case each of the above listed engagement, event or activity would have been a PROJECT in its own right and would have involved serious mental as well as physical activity on your part.

Thus, knowingly or unknowingly all of us are engaging in projects, albeit small, in course of our day-to-day lives and therefore almost all of us are unknowingly fully familiar with handling projects successfully. One example can be inviting two families for dinner at home. We do not require reading text books and reference books for successful execution of these ever-recurring projects that we unconsciously handle days after days and perpetually, almost always with success and satisfaction.

However, when it comes to handing an industrial project the job appears mountainous and the first thing that comes to mind of most of us is the doubt about our own capability to handle it successfully. It would be worthwhile to ponder a while to determine as to what suddenly goes wrong when our responsibility shifts from handling a personal domestic project to handing an industrial project.

The causative factor is want of the required Level of Fitness for efficient and successful handling of industrial projects. In the following we will take a look at what fitness aspects are needed at various levels.

## *Level of fitness required of the person who is to handle a project*

The very expression 'fitness' is generally considered to be a matter pertaining to physical health and strength. No doubt, physical fitness is the first requisite but equally important other aspects of fitness are generally not even recognized and taken care of.

Let us consider the following list of various aspects of fitness and try to find the ones that have no relevance in connection with efficient and successful handling of industrial projects.

- Physical Fitness: Good health,
- Mental Fitness: Good Memory, Discretion and reasoning,
- Psychological Fitness-1: Ownership towards the project,
- Psychological Fitness-2: Will to accept the challenge,
- Psychological Fitness-3: Team Spirit & Brainstorming,
- Psychological Fitness-4: Confidence,
- Mental Fitness - Patience,
- Integrity and Honesty,
- Having Faith (nothing religious here).

It definitely is debatable as to whether or not the above attributes qualify for being literally included under the term 'fitness'. However, the very purpose of this book is to deal with the subject of efficient and successful handling of industrial projects in totality rather than indulging in a philosophical discussion about the subject.

Out of the above attributes of a person's 'fitness' for handling industrial projects, the first two are of a nature where medical doctors and psychologists should best be consulted. Perhaps there may not exist any history or instance pertaining to a candidate being considered for handling industrial project having been examined on the lines to determine his/her fitness on the basis of attributes other than the first two. Therefore, in the following let us examine the remaining seven aspects of project management.

## *Ownership of the project*

It is the doubt in our minds about our own capability to handle industrial project that is one of the root causes of start of disaster in execution of the project. The reason for a doubt arising about his/her own capability is 'lack of ownership'. So far as capability of any person to handle any matter is concerned, it is the lack of ownership that is the principal causative factor of start of tendency to become a spectator rather than being a performer. Imagine a mother observing a snake approaching her toddler. Will the mother, in such a case, think twice and hesitate before running at her best to quickly lift away the baby? No, she will not and this reaction will be spontaneous without anyone having to prompt her to do so. The reason is ownership of the baby that a mother naturally has.

Coming back to industrial project, what really happens is that in the team meant for execution of the project there are two kinds of people – those who own the project (or parts of project that they are entrusted with responsibility of) and those who don't. It is those who don't own the project who constitute the stumbling blocks in the path of its progress. Just the feeling of ownership makes all the difference. A person who owns the project (or a part of it – for example a welder can at the most own the work where the welding is required to be done) automatically has his/her brain working with open 'eyes of wisdom' (in Sanskrit expressed as "Jnan Chakshu" which literally means 'Eyes of Wisdom'). Such a person is curious, inquisitive and imaginative always making in his/her mind a picture of what must be going on at the work site. In doing so, he/she is able to also conceive what should not be allowed to go on, for sake of safety or added cost or damage to equipment or time over-run. This type of vision of the person coming through eyes of wisdom lets him/her to often gather details of the place of work or piece of work that could have gone wrong, even without having to physically move around. In contrast, a person who doesn't have eyes of wisdom roams around the site aimlessly and often (as a matter of chance) doesn't reach the exact location where something is really going wrong or about to wrong or passes close to such a location without realizing what is about to go wrong.

At times, such perception or noticing through eyes of wisdom comes even with eyes closed as the person concerned is able to chart out a trajectory of what he/she has, some time back, actually seen and such perception leads the person to take advance precautionary measures even before any undesirable incident can occur. It is the perception, rather than glancing or looking that keeps the person's sub cautious brain on the job so that he/she is able to preempt undesirable occurrences and incidents. Such incidents include mishaps, accidents, damages, wrong methods being adopted, wrong material being used ... the list can be long.

Therefore, the most important prerequisite for a person to succeed in execution of a project of any kind is to own it in entirety and without exceptions. Such ownership is needed at all levels and it is the responsibility of the manager of a project to motivate all other team members to become owners of their respective parts of the project. Unless the project manager is able to first himself/ herself own the entire project and to motivate his/her team members to also equally and whole-heartedly own their respective portion of the project, it may generally be difficult to achieve 100% success in execution of the project concerned.

It may sound unrealistic but the fact is that seeds for the problem of lack of ownership are sown at the very level of Owner of the Project. This happens unknowingly as the owner looks at the Project only in terms of money – expenses versus profit. Because of this ultimately a commercial person (or a technical person with commerce and /or finance as the specialization) incidentally or consequently becomes the head of project management team.

## *Will and determination to accept the challenge*

We all know merit of the famous English proverb, "where there's a will, there's a way". The very first requisite for any person to be able to apply himself/herself to any matter, whatsoever, is his/her will and the greater the level of will the greater automatically becomes his/her intensity of focus of energies – physical as well as mental – towards achieving that which is being willed by him/her.

This is why the term 'will power' has gained popularity and the biggest strength of us humans is our will power. History is replete with instances of wars between two countries, one of whom was lesser equipped with military, navy and air force equipment while the other was equipped with state of the art armament, ships and fighter planes. Even then the country that was lesser equipped won the war – the victory having been brought about by 'will' of the fighting forces of the seemingly ill-equipped country. Quite obviously such 'will' of the fighting forces is in such cases a result of motivation and moral infused by the commander or general.

## *Team Spirit and Brainstorming*

A project will be executed fruitfully and smoothly only if project manager creates a Team Environment. Basically, creating a team environment is not very easy. To effectively build teams, it is important to remember that:

- Teamwork is generally based on a company's culture. Companies that encourage open, honest communication and foster employee interaction are in a better position to have good teamwork among employees.
- Team spirit comes from the top. Building effective

teams with the right attitude emanates from the highest levels of an organization. Only by flattening the traditional organizational pyramid can one expect to instill
- People must fit the culture. Some people are team players and some aren't. It's partly a question of personality and partly a matter of training. One person in the team with the wrong attitude can undermine the effort of the entire team. Hiring only people with the right traits for teamwork is crucial in building effective teams.
- However, a project manager should not feel helpless if his company is still following the traditional organizational pyramid. To a great extent a project manager in himself/ herself reaches in a position where he can try to inculcate a better working environment within his/her team.
- Key to successfully developing creative team spirit among members of project team lies in getting everybody on board so as to form a Team. Teams are not born, they are built. Here's how to go about building an effective team where everyone considers it his or her job to pitch in and contribute to the overall team effort.
    - Start at the beginning – chose people who fit into the desired team culture. Use interviews and earlier assessments well to determine a person's personality and traits. Look for cooperation and a willingness to listen in order to determine whether an individual can work well in a team environment.
    - Let everyone know where you're going. This means making project team members part of the strategic planning process and making sure they understand the goals. It's important that every team member buys into the plan.
    - Hold the team accountable for results. Establish clear team and individual expectations. Make sure everyone understands that it's the result of the team's

work that ultimately counts. Base your reward system on the team's ability to achieve its goals. Provide individual recognition based on contribution to the team effort.

If project manager is able to develop team spirit among project execution team, the next important step is to introduce process of Creative Team Brainstorming, advantage of which is that it improves critical thinking and problem-solving skills of members of project execution team as individuals and as a team. Team members often feel more open to bouncing ideas off one another and seeking advice on individual projects when creative team brainstorming is an important part of the work process.

## *Confidence*

Here confidence encompasses (i) self-confidence, (ii) confidence in colleagues and subordinate and (iii) confidence in top management of the organization.

The concept of self-confidence is commonly used as self-assurance in one's personal judgment, ability, power, etc. One increases self-confidence from experiences of having mastered any one or more particular activities. It is a positive belief that in the future one can generally accomplish what one wishes to do. Self-confidence is not the same as self-esteem, which is an evaluation of one's own worth, whereas self-confidence is more specifically trust in one's ability to achieve some goal, which is somewhat similar to generalization of self-efficacy.

Self-efficacy refers to an individual's belief in his or her capacity to execute behaviors (including actions) necessary to produce specific performance attainments. Here I would like to emphasize the need to distinguish between self-confidence as a generalized personality characteristic, and self-confidence with respect to a specific task, ability or challenge (i.e. self-efficacy). Self-confidence typically and loosely refers to general self-confidence. This is different from self-efficacy, which stems (from personal experience and abilities) as a "belief in one's ability to succeed in specific situations or accomplish a task and therefore is the term that more accurately refers to specific self-confidence. Psychologists have long noted that a person can possess self-confidence that he or she can complete a specific task (self-efficacy) (e.g. cook a good meal or write a good novel) even though they may lack general self-confidence, or conversely be self-confident though they lack the self-efficacy to achieve a particular task (e.g. write a novel). These two types of self-confidence are, however, correlated with each other, and for this reason can be easily conflated (sometimes mistakenly)

Lack of confidence results in the proverbial shaky legs and no challenging journey can ever be completed with shaky legs. Confidence comes from deep understanding of what one is required to do.

## *Patience*

Patience (or forbearing) is the state of endurance under difficult circumstances, which can mean persevering in the face of delay or provocation without acting on negative annoyance/anger; or exhibiting forbearance when under strain, especially when faced with longer-term difficulties. Patience is the level of endurance one can have before negativity. It is also used to refer to the character trait of being steadfast. Antonyms of patience include hastiness and impetuousness.

At times, despite the best efforts, there goes something wrong and in big projects such instances are reality rather than exception. No matter what has gone wrong, what gives real strength to the person responsible for taking the appropriate corrective steps is his/her patience in keenly analyzing the situation and finding way out to proceed further with as less as possible an impact on the project concerned. It is patience that imparts the required level of hardihood and strength to the project manager to face the circumstances with full presence of mind and successfully.

Lack of patience gives birth to anger and anger leads to confusion, loss of memory and resultant disaster.

The irony is that if a person at senior level loses patience, the loss permeates very fast down below to persons at lower levels and this results in shakiness in the entire organization that ultimately results in collapse of the structure of management which ultimately results in a disastrous failure of the project concerned.

## *Honesty*

Honesty refers to a facet of moral character and connotes positive and virtuous attributes such as integrity, truthfulness, straightforwardness, including straightforwardness of conduct, along with the absence of lying, cheating, theft, etc. Honesty also involves being trustworthy, loyal, fair, and sincere.

However, fact of life is that excess of anything is undesirable. Too much honesty might be seen as undisciplined openness. For example, individuals may be perceived as being "too honest" if they honestly express the negative opinions of others, either without having been asked their opinion, or having been asked in a circumstance where the response would be trivial.

Thus, in context of virtues of a successful project manager, honesty does not just mean, "never ever tell a lie under any circumstance"; it also is not just the virtue of "always telling the truth, no matter what the consequence". There can be no doubt that these virtues are important part of being a good person or human but the fact remains that being truthful only does not suffice for being honest.

In the present context honesty means never faking reality in order to gain a value. It is the virtue of refusing to pretend that facts are other than what they really they are. Here the terms 'faking' and 'pretend' actually pertain to a situation where a person either 'fakes' or 'pretends' to himself/herself and in so doing allows the project to take a course which ultimately proves to be harmful, often leading to situations of no return and resulting in dooming the project.

Such facts may even pertain to one's own qualities or competence. In order that a person can aspire to be a successful project manager it is necessary that the person honestly knows and acknowledges to himself exactly what he lacks in order to succeed and then acts positively and relentlessly to fill such vacuums caused by his/her own lack of some attribute that is required but is missing. When a person is thus honest, he/she makes efforts to get over the situation by borrowing help from some real expert exactly when required and thus is able to save some error or mistake or lack of something that would have crept in and resulted in some catastrophe harming the project badly.

## Having Faith

In the context of project management, Faith does not mean anything religious. There is no specific book of guidelines that defines "having faith". Having faith means whatever you want it to mean, but it does mean having an understanding of the ways of the world and letting things be. If you tell someone to have faith, you are encouraging him/her to stay confident.

Actually, having faith combines the above-described attributes of confidence, patience and honesty and will and determination to accept challenge results in a combination that ultimately leads to success.

Also, relevant to be added on the appropriate and essential attributes for a global project management candidates, which few educational programs emphasize, is the importance of understanding cultural issues. Introducing imported skills and management techniques always raises the possibility of importing new values and behaviours, which may be entirely foreign to and inconsistent with the values and culture of a country. Global topics that are useful for engineers and project managers to learn about include courses in language, world cultures, global business strategies, world history, etc. Two other areas that are useful when working overseas are cultural anthropology and archaeology because they provide a historical and ancient perspective on the development of other cultures.

One anomaly that is created by educational systems occurs when engineers and constructors leave developing countries to study in industrialized countries, where educational systems emphasize technology that is appropriate to highly developed countries or western cultures. Another problem created by educational systems is that some of the western-educated engineers and constructors, when working in key government positions developing new projects, usually disregard indigenous social and cultural aspects of designs. This could result in projects where there is no local expertise for competent operation of projects, or parts and services are not available for maintaining projects. In some situations, projects may be abandoned because the native population reverts to their traditional way of doing what the project tries to automate for them.

All these remarks only emphasize how difficult it is to have project managers who have appropriate backgrounds in all of the areas of expertise required to succeed managing projects in global arena.

# PROJECT STAKEHOLDERS AND THEIR ROLE IN SUCCESS

There is a saying to the effect that it is impossible for anyone to see the heavens without dying. Accordingly, with reference to execution of engineering projects, it is extremely important that each stakeholder in any project understands this and acts relentlessly, to the best of his/her might and in a manner befitting to his/her role in the project, to drive the project to success.

Stakeholders are individuals who are somehow influenced by or are influencing the project, and their interest may be positive or negative. Some belong to both categories. Knowing the project's various stakeholders and their respective expectations is of the vital importance for the survival of the project manager. Identifying and categorizing them will facilitate dealing and communicating with them later on. For example:.

- Core stakeholders: prime movers and/or decision makers. Usually the same as the project organization.

- Primary stakeholders: those who are particularly affected by, and would like to influence, the project.
- Secondary stakeholders: those with relatively low interest and as such will probably not influence the project.

For some stakeholders, it is sometimes more important how a project manager behaves than what he or she is actually producing. The project manager is likely to be assessed on a number of 'soft' criteria. Try to learn what matters the most to each stakeholder in order to ensure that the project's most important stakeholders have expectations in line with the project's goals.

Generally, the stakeholders in any project are:
- Developer (Owner or Promoter or Entrepreneur) - person or company or corporation),
- Financer or Debtor,
- Principal Contractor or Contractors (in case total project is divided between several contractors),
- Sub-Contractors,
- Suppliers,
- People or community affected by the project,
- Regulatory Authorities,
- Beneficiaries of the Project.

Unless each of the above listed stakeholders truly owns it and acts promptly and relentlessly to sail the project to success, it may rather be difficult for a project to really succeed.

While on the subject it would be in order to give a thought to what is meant by 'success' of a Project. In a broad sense, it means achieving project objectives within schedule and within budget, to satisfy the stakeholder and learn from experience. Brainstorming is likely to lead to the following:
- Upon being declared ready for testing, preliminary operations, commissioning and proving of performance

guarantees
- Should be complete not only mechanically or technically but also must be compliant (or ready for being proven to be compliant) of all rules and regulations and stipulations required of it for being fit for long term commercial operation,
- Should have received the permits, clearances and approvals stipulated to be available for being put into initial live operations (examples are approvals from Electrical Inspector, Boiler Inspector, Explosives Inspector and the like as may be applicable).
- Coming into successful commercial operation of the project well in time and without cost and/or time overruns,
- Achieving the above without,
  - Disruptions, except for force majeure events,
  - Spoiling environment,
  - Creation of nuisance for others,
  - Any person suffering from any major or minor health related incident or mishap,
  - Any plant or component or machinery or structure or material getting damaged during (i) transit, (ii) storage, (iii) handling, (iv) testing, (v) commissioning and trail operations and (vi) initial commercial operations.

## *Role of Owner or Promoter*

**Structure of Project management organization**

Owner or Promoter of a project acts through a set of its employees or officers of different levels of hierarchy for taking care of its interests in all possible ways. These employees or officers may be classified as below.
- Finance and/or Commerce managers,
- Administration managers,

- Engineering or Technical managers,

Depending upon nature of the project there can be a plurality of Engineering or Technical Managers. Some examples of nature of project are:
- Engineering Maintenance,
- Engineering Operations,
- Chemical Engineering,
- Metallurgical Engineering,
- Civil Engineering,
- Textile Engineering, etc.

Success of project depends upon proper coordination between these managers, as left to themselves their prime concern becomes hierarchy in the system rather than their contribution to the project. Each one is more concerned to impose him/herself on the others and to prove to all that his position is most important. In doing so these managers forget their main concern or their main responsibility of ensuring successful execution of the project.

Therefore, it is in the interest of the owner to put in place a proper mechanism for ensuring that the above-described situation doesn't arise. One effective method is to implement Flat Structure of organization. Writing much more than this would mean switching over to HRM (Human Resource Management) – which is not the subject matter of this book and therefore we put this matter to rest.

## Selection of Consultants of the required caliber

Owner of project appoints Consultant because (a) the financial institutions (FI) lending funds insist it in order to safeguard the amount of lending and (b) owner expects that the consultant would take care of owner's interests in ensuring that the contractor does the required work and delivers the required equipment commensurate with the required level of state of the art.

Often the FI proposes name of consultant of its choice. If this is the case, it is advisable that the owner properly evaluates the consultant proposed by the FI and ensures due competence of such consultant to serve the owner's purpose and expectations. For evaluation of the consultant it is advisable to ensure the following.
- The consultant has already successfully rendered consultancy services for at least two comparable or bigger projects and
- Owners of these projects are fully satisfied with the services rendered by the consultant and at the moment of evaluation of consultant these earlier executed projects are commercially adequately profitable.

However, some further due diligence towards evaluation of the consultant is also advisable. The reason is that consultancy has become a namesake business and a consultant doesn't generally tailor make its specifications and other documents for each specific project, rather the entire documentation of most of the consultants are run-of-the mill type, mass produced and made project specific only through copy-and-paste. This type of approach is not advisable because main equipment manufacturers go on continuously innovating in order to remain competitive by way of better performance as compared with competitors and often reduction of cost.

## *Most important aspect: Selection of Land Plot for the Project*

Selection of Land Plot for a project in India or elsewhere is not always made on the basis of real scientific assessment of suitability of it for the particular project. More often than not, projects are set up based on whims and fancies of politicians as powerful ones manage to get certain areas allocated for setting up of various projects, in the name of 'development' of the area or the state and attractive concessions are thrown at the entrepreneurs so as to attract investment for setting up industries in such areas. For example, in a certain state named 'XYZ' suddenly an area is declared as having been allocated for setting up industries under 'XYZIDC' (i.e., State Industrial Development Corporation (SIDC) for the state 'XYZ' – such SIDCs are meant to serve as a single window problem solvers and administrators of the areas so declared to have been allocated for setting up of industries. Following are typical examples of some of the main objectives of SIDCs:

- To stimulate the growth of industries in the small/medium scale sector,
- To provide infrastructure facilities like roads, drainage, electricity, water supply, etc. is one of the primary objective of SIDCO.
- To Promote industrial estates which will provide industrial sheds of different sizes with all basic infrastructure facilities.
- To Provide technical assistance through training facilities to the entrepreneurs.
- To Promote skilled labor through the setting up of Industrial Training Institutes.

More details can be seen from websites of various SIDCs.

It is only the developer of a project who really knows why a particular project is being undertaken. It is just possible that the effect of selection of a plot of land that is really not suitable for the purpose may be of no consequence whatsoever. However, irrespective of intention of the developer, it is necessary for the very purpose of this book to explain the matter of selection of Land Plot for a project in truthful details.

In particular, it is important to realize that despite there being a SIDC in picture there is no guarantee that any particular

Plot of Land, chosen without very careful considerations, shall prove to be suitable for the normally conceivable purpose of the project.

It is the job of a group of experts to adjudge the suitability of any particular plot of land for setting up a particular industry. The plot that may be suitable for setting up a rice mill may not at all be suitable for setting up of a thermal power plant or a wind based power plant. Following are the basic criteria for any particular plot of land to be suitable, in order of importance.

- Adequacy of land area,
- Connectivity of the proposed land for transport of goods and storage areas, both those required as input for the project as well as the raw material and product, including accessibilities in general,
- Availability of adequate quantity of water of the required quality at affordable cost,
- Reasonable proximity with infrastructures premises, including hospital, fire department, accommodation & recreation facilities.

# PROJECT CONTRACTING

Except for the projects involving negligibly small cost, all cost intensive projects are executed under contract. Therefore, Contracting is really the most important aspect of projects. Each of the stakeholders of a project has different interest and likes the contract to be framed so as to serve its interests.

For example, the very business of contracting means a near-zero investment affair with near infinity ROI (Return on Investment) whereas for any owner (entrepreneur or promoter) the cost of project is in itself the investment because projects by themselves are not the ultimate aim of a promoter whose sole interest lies in earnings from the project in future. Therefore, while the owner would like the cost to be least and cash flow such that interest during construction is as low as possible, the contractor would like the cost to be as high as possible and front loaded to maximum extent so that the payments received from owner form a good enough buffer for managing rest of the project at zero cost to the contractor.

In the following is a list (not exhaustive) of various important financial aspects and the interests of various parties involved in a contract from their point of view:

- **Cost**: the owner's/developer's interest is in minimizing cost but from the contractor's point of view the higher the cost the better;
- **Cash Flow**: from the owner's/developer's point of view the Interest During Construction (IDC) being an inseparable part of total project cost should be the least possible. To this end Cash Flow should be the minimum possible from beginning to end and maximum Cash Flow should occur at end of the project. On the other hand the contractor prefers to front loaded as much as possible. Since contractor generally does not spend any money of its own and starts its work utilizing the money received as 'Advance Payment' from the owner and thereafter from periodic 'progress payments', its intention is always to obtain maximum amount right from the inception.

Since commerce and economics are not the key subject matters of this book the matters pertaining to cost and finance are not being further elaborated.

In all types of industrial projects there is one more very important player – the Technical or Engineering Consultant (the Consulting Engineer). More in details follows about this equally important entity.

## Nature of Contract and Number of Contractors Responsible for Project

There are several ways of execution of a project:

- **In-house Execution** – wherein no major contractor is involved. The team of the owner executes all work in such a case, with exception in some case being the work of technical consultant or 'Consulting Engineer'. Construction Machinery in such cases is either hired or procured by the owner depending upon size and complexity of project. In this method of execution of project there is no concept of any overall guarantee (each of OEMs guarantee performance of their respective equipment or machinery or plant and this guarantee is figurative and subject to being demonstrated to owner by respective OEM on completion of project). Similarly, in this method of execution of project the concept of warranties too is a matter for each of OEM and parties that participate in project for its execution. The subject of guarantees and warranties would not be discussed any further since these matters are more of commercial nature than being of technical nature and are not directly linked with execution of project.
- **Works Contract or Free Issue Contract** – wherein all machinery and material, including construction material, is supplied by the owner at its own cost without any costs towards these having to be met by the contractor. There can be more than one Works Contracts and therefore several contractors. With such contract/(s) the owner undertakes the responsibility of procurement and free of cost supply of either all project machinery, equipment and construction materials or only the major project machinery, interconnecting equipment,

the balance being in scope of the contractor.

Subcontracting brings with it a whole host of issues that are nonexistent when self-performing. Similar to the effect that multiple projects in the area have on the supply of labor, subcontractors also can get stretched too thin. Some will accept the work and then not be able to perform, while others will accept the work, do an unsatisfactory job, and ultimately cost the project more time and money than if the work were self-performed.

Another aspect also to be consider into the decision making of contractor candidates and estimating process are the union versus non-union filiation status; possible differing pay scales; distance from the job sites; and per diem for travelers. Such disparities between new contractor incomers and the contractors already at site may create disturbances, and therefore requires also attention. One of them is time and materials (T&M) in which the employer agrees to pay the contractor based upon the time spent by the contractor's employees and subcontractor employees to perform the work, and for materials used in the construction (plus the contractor's markup on the materials used), no matter how much work is required to complete construction. Time and Materials is generally used in projects in which it is not possible to accurately estimate the size of the project, or when it is expected that the project requirements would most likely change.

There is no hard and fast rule about such contracts and scope of works and supplies can have a large number of variations. In this modality, the estimator develops a set of unit rates for labor, tools, equipment, supervision, and so forth. These rates are then tabulated and made part of the contract, along with a stipulation that any material purchases would be passed on to the owner at cost plus some specific markup.

In this method of execution of project also there is no concept of any overall guarantee (each of OEMs guarantee performance of their respective equipment or machinery or plant and this guarantee is figurative and subject to being demonstrated to owner by respective OEM on completion of project). Similarly, in this method of execution of project the concept of warranties too is a matter for each of OEM and parties that participate in project for its execution. The subject of guarantees and warranties would not be discussed any further since these matters are more of commercial nature than being of technical nature and are not directly linked with execution of project.

- **EPC (Engineering, Procurement & Construction) Contract** (also termed as Turn-key Contract or Lump Sum Turn Key (LSTK) Contract – wherein owner selects a party of very high reputation (generally a multinational company) and executes a contract whereby the EPC Contractor takes total responsibility of executing all the engineering and technical studies and design and engineering, manufacturing (of generally the main most important plant and machinery), procurement of miscellaneous plant and machinery, fitments, hardware, consumables, building materials and all that can be needed to complete the project. In fact, the least risky method of contracting for the installation of major power plant components and its auxiliary systems and facilities is for the owner or general contractor (GC) to put the responsibility of the work and performance of the equipment and the plant as a whole on the designer or supplier of that equipment (OEM). Although there are many instances when this may not be cost-effective, from the perspective of the construction portion of the project, this type of contracting arrangement generally provides to the owner the longest-term warranty that the equipment will perform as specified. When projects are designed by the OEM to install its own equipment, productivity frequently improves, considering its expertise accumulated in several similar other projects.

In this type of project execution, the matters of guarantee and warranty form important aspect and project manager has to carry out performance test/(s) for demonstrating to owner that performance as guaranteed in contract is really being achieved. Also, warranties become responsibility of the party entrusted with execution of project. Warranties do not require any specific actions on part of project manager and come into picture in due course if and when any questions arise as to a warranty breach. Such matters do not form a part of project execution and therefore are not being taken up in this book any further.

When the higher cost of contracting to an OEM is perceived as not adding value to the project, the owner or GC may elect to contract to a local third party. This would usually be at expense of decoupling the warranty of the equipment from the warranty of the installation work. This decision has positive and negative implications. On the positive side, local third-party contractors frequently have experience working in that particular area, meaning they are familiar with the labor, unions, and even may have "sweetheart" deals to hand-pick qualified craftsmen for the work. Also, special arrangements with tool and equipment suppliers. Regarding the downsides, besides not having free access to information from OEM, there may be the cost of OEM representation on-site, as required by the OEM to validate the work done by the third-party contractor. This can be costly because the OEM is usually the one determining how long this representative is required to be on site. Other costs to be considered are the possibility of material damage in transit and repair to equipment damaged during installation.

Without the OEM taking full responsibility for these issues, the owner or GC may encounter lengthy delays and costs in trying to resolve them.

Whether it is the owner, the GC, or the non-OEM third party that is providing the labor, one of them is doing the actual hiring and firing. There are different potential sources of direct hiring labors, namely: union labor, non-union labor, or via labor brokers.

When considering union labor hiring sources, in case the plant and contractors are already signatories' unions legally binding agreements, there are fixed labor rates from each of the signatory unions (electricians, pipefitters, etc.) for a defined period of time. Sometimes there will be adjustment factors to be considered. Irrespective of the labor contracts in place between unions and the contractors, the owners of some major projects use the amount of the upcoming works as leverage to obtain special union rates. This is commonly called a PLA (project labor agreement). A PLA generally locks in the craft labor wages for a total duration of the project and of course obliges all project participants to use unionized labor.

Unlike that case of unionized labor, when a contractor uses non-unionized labors (open-shop), there is flexibility to pay whatever the labor market will bear and also freedom to be innovative with the benefits package (e.g. bonuses, productivity incentives, flexible shift and labor hours, less stringent hiring and firing requirements, etc). This flexibility, however, creates more uncertainties on the final cost of these resources. Notwithstanding this, it is a good opportunity to have lower labor costs, since the union wage rates include more than just workers' wages and benefits, but also the cost of the union infrastructure.

The last source of hiring is through the labor brokers. It normally happens when there are no local available qualified workers in the quantities required within union or in the market. Labor brokers usually require a signed contract with the contractor, fixing the hourly or daily rates and fringe benefits for the duration of the project. Sometimes, using a labor broker is more cost-effective than the standard union or non-union approach. Pricing craft labor from a labor broker can result in a lower overall cost, although not always immediately apparent, moreover by the fact that this approach relieves the contractor, GC, owner the responsibility to have a structure of such administrative payroll services on site, at least for these workers.

Note: Post completion of execution and commissioning under EPC there can be further involvement of the party responsible for EPC execution in terms of BOO (Build, Own and Operate) or BOOM (Build, Own, Operate & Maintain) or BOOT (Build, Own, Operate and Transfer) agreements, if included in contract. Such contracts, more often than not pertain to power projects – either thermal power plant or hydroelectric power plant. Other industries, even as an exception, are not known to have indulged in these methods of project contracting. So far as execution of project under EPC contract is concerned there is hardly any difference in execution irrespective of whether the contracts ends after handing over of facility after EPC or there has to further involvement in terms of BOO, BOOM or BOOT agreement. Differences between these are totally commercial and operation & maintenance of any industry post successful completion of execution of project does not form a part of project execution. Therefore, these topics are not covered in this book.

For enabling smooth execution under EPC, a lot of inputs generally are required to be provided by the owner. For example, the following:

- Adequate area of land free of all encumbrances over which project is supposed to be setup. However, someone has to make sure that the land is not prone to flooding (also meaning thereby that there is sufficient enough natural drainage of storm water to somewhere for which there would later be no legal objections), does not have mineral or archaeological reserves, is not prone to disastrous earthquakes, is reachable with good enough roads and railways for enabling fast and safe transport of all plant, machinery and material – and this responsibility should better be taken by the owner although there are no hard and fast rules,

- Source of Raw Material/(s) and necessary commercial tie-up for supply thereof including the required proper transport,
- Taker of the product,
- Secure and approved (by the competent authority) area for disposal of industrial waste and right of way for its transport or carriage, as may be applicable according to various rules,
- Environmental Clearances as may be necessary under law of the land,
- Electricity and water for construction work,
- Land for proper housing or camping of manpower to be engaged for execution of project and availability of potable water, sewage disposal system and other basic necessities for sustaining life of the construction workforce,
- Facilities of medical treatment for the workforce – particularly good enough for serious emergencies,
- Tie-up with local Fire Fighting Arrangement,
- Water and Power for start-up and commissioning of the project,
- All the required Permits and Clearances.

In case of In-house execution of a project all risks and responsibilities lie with the project owner and the matters pertaining to responsibility have no ambiguity.

However, in cases of Works/ Free Issue Contract or EPC Contract, there must be unambiguous clarity about which party is supposed to be responsible for Sub-soil Risks. Here the term 'Sub-soil Risk' is not only limited to the nature of sub-soil of the site but also risks related to questions that may arise in the following cases:

- If there is discovered some hidden or earlier unknown big void underground such that may be posing sufficient enough risk to civil construction that may be considered as reliable enough for long term trouble free functioning of the plant and machinery of the project;
- If a mineral source of some kind that was earlier unknown is discovered during execution of the project;
- If soil strata underground below a certain depth is found to be such as may be requiring some high cost civil work that was not originally foreseen or envisaged.

Of the above three the first two are matters about which only a consensus is to be reached between the parties. How such a consensus will be reached is beyond the scope of this book.

However, for the third case namely sub-risk involving discovery of some unforeseen soil strata requiring high cost civil work, underground at a certain depth is concerned, the best way to avoid such a situation is to first prepare detailed layout of the project detailing therein locations of all heavy loads, in particular those involving machinery, that would be likely to continuously produce vibrations leading to sinking of foundations. Next step is to judiciously decide large enough plurality of locations for carrying out soil load bearing capacity up to a certain depth and also in-depth soil investigations through drilling of bore holes to extract soil cores up to large enough depth until hard enough strata capable of supporting loads is reached or strata suitable for friction piles good enough to support heavy and vibrating loads is found. This should be done under guidance of some reputed expert of soil mechanics and expenses incurred towards it should be factored well in advance.

The important thing is for contractors to at least be aware of the risks involved and that the common law places the responsibility for ascertaining the ground conditions and being able to carry out and complete the works in the light of those conditions, on the contractor. Contractors who are not prepared to take these risks must ensure that the contract is amended so that their risk-taking is reduced to an acceptable level by means of having been shared by the owner and in such a case this fact should clearly and definitely be described in the contract so that no dispute on this matter arises at a later date.

## Clarity about Deliverables and Scope related to Deliverables

Each of the contracts of a project must be drawn properly and covering all the aspects that would come into play during execution, commissioning, putting into commercial operation and final taking over by the Owner.

No doubt, in general lot of technical points are defined in contracts in good enough - and sometimes more than enough – details. Commercial and Legal part of contracts similarly contain good enough details of mutual give and take between the parties involved in each contract.

However, when it comes to actual execution of a project there occur times when execution reaches defined as well as unexpected milestones where no clarity exists as to scope of deliverables and/or scope of work.

In general, the term 'Deliverables' covers two (2) types namely (i) Project Deliverable and (ii) Process Deliverable. While each of these varied deliverables serves a specific purpose within the project management process, they also work hand-in-hand, and you can't overlook one for the other.

Project Deliverables result in Outcomes of any Project. By definition, projects are initiated to produce specific, unique outcomes based on specific, unique needs (that's how they differ from day-to-day operations). But in real life when more than one party are involved the implied meaning doesn't serve the real purpose as disputes arise between the parties as to whose Deliverable a particular entity really is as a Scope. Therefore, it becomes necessary that that purpose and need of 'deliverable' is expressed in some tangible form, whether it's a product, a process, a plan, a policy, or some other outcome. And what is defines in such tangible form becomes the deliverable.

Process Deliverables are the means by which projects are planned, managed and executed. In fact, for any given project, it may take multiple process deliverables to produce the type of timely, high quality project deliverables that are expected and required. In simple terms, if the project deliverable is the destination, the process deliverable is the roadmap used to get there. And every project needs it's own roadmap.

Nonetheless, hereunder, I am listing as many of deliverables as I can within the limitation of lack of specifics. Therefore, the following is a guesstimated list of deliverables:

## Examples of Project Deliverables

### Land area for project

There can be two alternative scenarios namely, execution of project (a) at a virgin site or (b) at a site already having a previous industry. Correspondingly, the actions required towards project management vary considerably, as described below.

(a) In a case where project is to be executed at a virgin:

- Post Land Acquisition there are two very important requirements in respect of development of the site and its land in order to make it possible to undertake necessary construction work at the site namely, (i) making site free from streams of storm water flowing in from surrounding land areas, if necessary by construction of a suitable garland canal good enough in capacity to divert the entire inflow of storm water from surrounding land areas.

    It is out of scope of this book to describe the entire process but just to give an approximate idea to the reader about what all can be needed to be done some basic steps are described below, and it is highly advisable to consult some civil engineering expert to study geological and geographical conditions of the site, rather than to do something based on guess work. The civil engineering expert should be expected to do the following.

(a) Carry out detailed study and analysis of Topographical Data given in Topographical Maps of the area obtained from Geological Survey, so as to determine overall natural storm water drainage of the area and vulnerability of the site to get flash-flooded by the inrush of storm water from surrounding highlands in cases of cloud burst (which occur rather frequently).

(b) Working out design of a garland canal around the entire project land capable of adequately diverting stormwater inflow from surrounding high lands, even in worst storm water incoming from sudden cloud burst of the worst kind statistically arrived at from past historical data, for depositing the storm water into some natural location such as river or sea.

- Making the project site suitable for working during rainy season – which in India stretches for three to four months, generally with two months of heavy downpour including cloud bursts – and requires ensuring that storm water precipitating from clouds is efficiently drained out to some system capable of adequately accepting the entire volume of such precipitating storm water inflow.
- Making top surface of soil at the site good enough for vehicular traffic, for construction purposes, in all seasons. This would need doing the appropriate depending upon types of the soil at the site. India has several types of soils at different locations in various states and it

is highly advisable to consult reputed experts, preferably premier engineering institutes (Indian Institutes of Technology) for this purpose. Appendix-1 gives a brief description of the type of soils in India.

(b) In a case where project is to be executed at a site already having a previous industrial installation:
- Making an unbiased assessment of the requirements in respect of conductivity of the top soil for trouble free vehicular movement during the worst rainy season conditions and taking actions as in case (a) above if needed.
- Enquiring from the concerned engineers and officers who are managing the existing industry about instances of minor and very temporary or major flooding of the industry site during any reason season ever since commissioning of the industry. However, while trying to obtain this information through talking and discussions, whatever, is told by the respective persons is to be taken with a pinch of salt since people are used to offering opinion subjectively and based upon their own person definition of what is normal and tolerable and what is not.

In this connection, I would like to describe an instance when a gas turbine based combined cycle power project was under construction on EPC basis by a company where I too was an employee. Although, I was not directly involved in the project, so far as its execution was concerned, the project manager concerned was a close friend and we used to meet socially. It so happened that my friend, the project manager, invited me for a visit to the project site just to show what all was going on and how he was handing the project. While moving around at the site with my friend we came to a location where there had been constructed a beautiful concrete pit of about 3 meters depth and rectangular floor of dimensions of about 5 x 8 meters and some pumps, piping –valves, power and control cables were being installed. I stopped at this location and started look at it with interest and with curiosity as I observed some things that did not look logical.

First abnormality that I noted was that there could have been an alternative design whereby the underground construction could have been avoided (all underground constructions for placement of machinery are risky and should be avoided unless there is absolute necessity).

The second abnormality that I noticed was that walls of the pit were protruding all around by about 150 mm or 200 mm above the grade level of the surrounding site (there was good enough rain cover, by way of nicely designed corrugated aluminum sheeting and a good enough hand railing of about 1500 mm height was also provided all around the top of pit walls).

A pit drain pump was also provided but its size gave me the impression that it was too insufficient in capacity to handle the incoming storm water, should there be any flooding of the area. I pointed this out to my friend but he said that the project was being executed on a site that is not virgin and the existing petrochemical plant has been successfully operational for last about a decade and a half with no report of flooding ever since its commissioning.

I couldn't say much but in my heart, I felt that the pit walls should have been protruding at least about 750 or 800 mm above the grade level. I did clearly mention this to my friend and frankly mentioned that if I was responsible for the project I would have firstly insisted on change in the engineering and design of the respective pumping system so as to eliminate the pit and if for some good reason putting the pumping system in a pit would, for some unlikely reasons, have been found to be impossible – I would have forced the engineering group to change pit design to make pit walls to protrude no lesser that 750 to 800 mm about the surround soil grade level.

Anyways my friend advised me not to bother too much about it and we both forgot the matter. It was about first week of July when I had visited the above described project site which was in the state of Gujarat located about 300 km north of Mumbai/ India and was near sea shore (about 5 km from sea). Rainy season was supposed to set in at the site around end July. On some date in end of July I came to know over a phone call from my friend, the project manager, that there had been a severe rain comprising a cloud burst for no lesser than 3 to 4 hours and entire site had got flooded as a result of which unexpectedly large amount of storm water had entered the pit and flooded it up to the top.

After rains subsided, flood water was pumped out from the pit, the machinery was taken out of the pit and dried out and serviced and civil engineering design of the pump pit was changed so as to raise protrusion of all its walls above the grade level by 700 mm. Subsequently, after completion of the said project I met my friend again and asked him as to what had given the design and engineering group the impression that protrusion of pit walls just by some 150 to 200 mm was sufficient. I was told that in the very first meeting with the customer for collection of site related data, it was confirmed by the customer that the site was totally free of flooding but after the disastrous flooding of pump pit, questioning by my friends with the workmen it was revealed that water logging of the site by about six to ten inches was frequent each rainy season and nobody was bothered about it because the petrochemical plant had nothing installed underground and the petrochemical plant by itself was having a plinth height of 600 mm above grade level.

I hope that my describing in the above in details does

serve purpose of bringing home the point that one should never blindly believe what is verbally informed during one-to-one talks or meetings about any matter that is technical in nature (and all matters, except the legal ones, pertaining to a project site, are strictly technical) and what should be believed are facts and figures and calculations made available as officially handed over authentic project related site data, that can fix contractual responsibility, of the party providing such data as and when it may become necessary. In absence of such authentic data, it is advisable to carry out due diligence to obtain scientifically correct data.

**A caution pertaining to a non-virgin project site**: I would like to mention here one very important caution that is often ignored. In the existing industry, there certainly would be metallic pipes, earthing-mat or grounding-mat and other metallic objects buried in soil. There is likelihood of there also being similar or other metallic objects to be buried in soil in the upcoming project. Chances are high of there being electrical conductivity above ground between the protruding parts of these pipes, earthing or grounding mats and other such buried metallic objects. This, if due care is not taken, may form formidable source of electrolytic corrosion for the sub-soil metallic pipes and other metallic objects of either the older industry or the upcoming industry or both. This occurrence of sub soil metallic objects needs deep study and necessary mitigation steps need to be taken to prevent it. Experts in project engineering and design group or at the end of engineering and design consultant or external experts may have to be involved in this matter so that ultimately the older industry and the upcoming one both remain free from subsoil electrolytic corrosion of metallic objects. Experts may include protective measures such as sacrificial anodes and/or electrical cathodic protection.

- Consents, approvals, permits and certifi-

cates related with land area for project and nature of project, for example approval from Pollution Control Authorities;
- Main plant and machinery;
- Auxiliary plant and machinery;
- Accessories of the above;
- Spare parts of plant and machinery;
- Cables, pipes, racks, hangers, supports, trenches, poles ... etc.;
- Buildings;
- Main and Auxiliary Roads;
- Laboratory Building/(s);
- Laboratory Equipment;
- Engineering of project including detailing to produce working drawings and tables.
- Approval/ commenting of engineering documents and drawings (by Owner and/or Consulting Engineer) as may be relevant of Project Deliverables, in terms of contract;
- Training Programs

The above is not a comprehensive list of Project Deliverables but I hope to have amply clarified the concept behind the term 'Project Deliverables' and leave it for the reader to amend and/or complete the list for his/her purpose because the process deliverables cannot be same for each and every project, rather the deliverables for each project are specific to that project.

An effective method for visualizing the project deliverables is to structure the work in a way that no imperative scope will be missed, which undoubtedly would bring about changes and additional work during the project. Also, this structuring will be the basis for making realistic schedule with estimates of durations and costs. In far too many projects the project managers have moved on directly to the activity planning and creating a nice looking, but useless schedule. During the structuring process, the project goals are broken down into smaller parts which can be illustrated in a hierarchical structure. This division can derive from sub-deliveries, areas of responsibilities, target groups, components or anything which is an important part of the project. The number of levels in the structure is determined by the need for detail descriptions of the contents in the separate parts of the project. If what is to be done is not clear about some part of the project, it is recommended to break it down to a lower level. The first level that is broken down to is called the main package, while the lowest level is called the work package and the whole structure is called Work Breakdown Structure (WBS). Note that the WBS does not show any link dependency nor sequence, it only shows the complete project deliverables in a structured way.

*Process Deliverables*

- Project site land development including grading, leveling, drainage planning and provision for temporary use and long-term use, boundary wall and approach road for the Project site land.
- Consents, approvals, certificates, permits and the like related with development and construction of project;
- Approval/ commenting of engineering documents and drawings (by Owner and/or Consulting Engineer) on a regular and recurring basis, as may be relevant to project and scopes decided and defined in contract;
- Carrying out of special tests such as Alkali Aggregate Reaction Test for ascertaining compatibility between the Portland Cement and stone aggregate proposed to be used for each specific project and carrying out extensive Soil Investigation through Bore Holes at close enough intervals in site where heavy and dynamic loads are expected to be encountered all through the life of project;
- Testing samples from each batch of construction materials including concrete reinforcement steel bars, bricks, steel joists etc. This is very important in India as suppliers tend to supply materials of quality poorer than specified and this they try to do randomly and repeatedly;
- Storage space and lay down areas;
- Tools and tackles, their repair, calibration if required, and replacement;
- Construction machinery;
- Power and/or fuel supply for construction machinery, including the responsibility of continuous availability of electrical power;
- Approach for construction machinery including right-of-way, particularly where crossing of pathway with railway lines, transmission lines, passages under bridges

- (adequacy of safe clearance in all directions);
- Ensuring adequacy of tests related to site and carrying out of further tests in case the available test results are deemed inadequate;
- Illumination of work site, letdown areas and passageways;
- Security of project site and limits of scope of security in the sense of extent of security as well as spread of areas of project land covered in scope of each entity (because in certain projects there may be more than one entity as contractors and even if there is only one contractor, there may be presence of owner in certain areas),
- Of equal or more importance is clarity of responsibility in this respect at interface points;
- Providing medical facilities, first aid and ambulance at site, meeting of cost of medical/ surgical treatment and hospitalization of workmen engaged at site as well as to third party person/(s) that may get involved in accidents by their own error or due to negligence – particularly where more than one contractor are involved,
- Payment of compensation to injured workmen as well as to third party person/(s);
- Payment of compensation towards damage of property, including that belonging to owner and/or other contractor/(s) or third party or parties;
- Responsibility towards persons aggrieved due to some aspects of a project (for example, noise from construction machinery or from use of explosives);
- (I hope to have amply clarified the term 'Process Deliverable' and am leaving it for the reader to amend and/or complete the list for his/her purpose).
- Ensuring strict compliance of all requisite Labour Laws and maintaining necessary records.
- Delivery of "Work" as well as "Process Fluid/(s)" and/or "Electrical Power" to others at interface points, in case there is more than one agency responsible for execution

of project, assuring due coordination among organizations.
- For example, another contractor may have been given contract to construct an electrical substation for providing power to the main project and providing potable water, at an interface point, for control room of the substation may be in scope of the main contractor. In such a case, the main contractor would have to deliver necessary pipe work up to the interface point and when connection has been completed at the interface point the main contractor would also have to deliver water of the specified quality as part of its deliverables. In this example if the contract is not clear about which of the two agencies would have to make proper physical connection of water pipes at interface point, there may arise a dispute at some stage.
- Commissioning of facility and carrying out Performance Test or Performance Guarantee (PG) Test, if applicable.

The above is not a comprehensive list of Process Deliverables but I hope to have amply clarified the concept behind the term 'Process Deliverables' and leave the topic here for the reader to amend and/or complete the list for his/her purpose because the process deliverables cannot be same for each and every project, rather the deliverables for each project are specific to that project.

## Legal Responsibility of Project Owner and/or Contractor/(s) towards Laborers

There are several ways of contracting and executing a project as explained above.

However, irrespective of the manner and location in which a project is executed respective Labour and Employment Laws are applicable equally. Compliance of all of these is basically responsibility of primary owner or developer of project, who should enforce the same for all chain of contractors and sub-contractors along the execution of the project. For convenience, and for the sake of avoiding overlapping responsibility, contracts do specify, in details about scope of deliverability of compliance thereof for duration of project. Nonetheless, the primary owner should better not forget that ultimate answerability and responsibility is vested with it in the eyes of the laws of the land, generally in all countries.

In the following are described the important Labour and Employment Laws of India for example those listed below (however, additions and alterations keep occurring from time to time and one needs to keep abreast with the latest developments as they occur).

**Labour and Employment Laws of India**

The Labour enactment in India, is divided into 5 broad categories, viz. Working Conditions, Industrial Relations, Wage, Welfare and Social Securities. The enactments are all based upon Constitution of India and the resolutions taken in ILO (International Labour Organisation) conventions from time to time.

Indian Labour law refers to laws regulating employment. There over fifty national laws and many more state-level laws.

Traditionally Indian Governments at federal and state level have sought to ensure a high degree of protection for workers through enforcement of labour laws.

While conforming to the essentials of the laws of contracts, a contract of employment must adhere also to the provisions of applicable labour laws and the rules contained under the Standing Orders of the establishment.

Indian Labour laws divide industry into two broad categories:

### 1. Factories Act

The Factories Act, 1948, regulates Factories in India. All Industrial establishments employing 10 or more persons and carrying manufacturing activities with the aid of power come within the definition of Factory. The act makes provisions for the health, safety, welfare, working hours and leave of workers in factories. The State Government through their 'Factory' inspectorates enforce the act. The act empowers the State Governments to frame rules, so that the local conditions prevailing in the State are appropriately reflected in the enforcement. The act puts special emphasis on welfare, health and safety of workers. The act is instrumental in strengthening the provisions relating to safety and health at work, providing for statutory health surveys, requiring appointment of safety officers, establishment of canteen, crèches, and welfare committees etc. in large factories.

The act also provides for specific safeguards against use and handling of hazardous substance by occupiers of factories and laying down of emergency standards and measures.

### 2. The Shops & Establishment Act

The Shops and Establishment Act is a state legislation act and each Indian state has framed its own rules for the Act. The object of this act is to provide statutory obligation and rights to employees and employers in the unauthorized sector of employment, i.e., shops and establishments. This act is applicable

to all persons employed in an establishment with or without wages, except the members of the employers' family. This act lays down the following rules:
- Working hours per day and week.
- Guidelines for spread-over, rest interval, opening and closing hours, closed days, national and religious holidays, overtime work.
- Employment of children, young persons and women.
- Rules for annual leave, maternity leave, sickness and casual leave, etc.
- Rules for employment and termination of service.

The main central laws dealing with labor issues in India are given below:

- **Minimum Wages Act 1948:** The Minimum Wages Act prescribes minimum wages for all employees in all establishments or working at home in certain employments specified in the schedule of the act. Central and State Governments revise minimum wages specified in the schedule from time to time. The Minimum Wages Act 1948 has classified workers as unskilled, semi-skilled, skilled; and highly skilled.

- **Industrial Employment (Standing orders) Act 1946:** The Industrial Employment Act requires employers in industrial establishments to clearly define the conditions of employment by issuing standing orders duly certified. Model standing orders issued under the act deal with classification of workmen, holidays, shifts, payment of wages, leaves, termination etc. Generally, the workers are classified as:
    - Apprentice/Trainee;
    - Casual;
    - Temporary;
    - Substitute;
    - Probationer;

- Permanent; and
- Fixed period employees

- **Payment of Wages Act 1936**: Under the Payment of Wages Act 1936 the following are the common obligations of the employer:
    - Every employer is primarily responsible for payment of wages to employees. The employer should fix the wage period (which may be per day, per week or per month) but in no case it should exceed one month;
    - Every employer should make timely payment of wages. If the employment of any person is being terminated, those wages should be paid within two days of the date of termination; and

  The employer should pay the wages in cash, i.e. in current coins or currency notes. However, wages may also be paid either by cheque or by crediting in employee's bank account after obtaining written consent.

- **Workmen's Compensation Act 1923**: The employer must pay compensation for an accident suffered by an employee during the course of employment and in accordance with the Act. The employer must submit a statement to the Commissioner (within 30 days of receiving the notice) giving the circumstances attending the death of a worker as result of an accident and indicating whether the employer is liable to deposit any compensation for the same. It should also submit an accident report to the Commissioner within seven days of the accident.

- **Industrial Disputes Act 1947**: The Industrial Disputes act 1947 provides for the investigation and settlement of industrial disputes in an industrial establishment relating to lockouts, layoffs, retrenchment etc. It provides the machinery for the reconciliation and adjudication of disputes or differences between the employees and the employers. Industrial undertaking includes an undertaking carrying any business, trade, manufacture etc.

The act lays down the conditions that shall be complied before the termination/retrenchment or layoff of a workman who has been in continuous service for not less than one year under an employer. The workman shall be given one month's notice in writing, indicating the reasons for retrenchment and the period of the notice that has expired or the workman has been paid, in lieu of such notice, wages for the period of the notice. The workman shall also be paid compensation equivalent to 15 days' average pay for each completed year of continuous service. A notice shall also be served on the appropriate government.

- **Employees Provident Funds and Miscellaneous Provisions Act 1952**: This Act seeks to ensure the financial security of the employees in an establishment by providing for a system of compulsory savings. The act provides for establishments of a contributory Provident Fund in which employees' contribution shall be at least equal to the contribution payable by the employer. Minimum contribution by the employees shall be 10-12% of the wages. This amount is payable to the employee after retirement and could also be withdrawn partly for certain specified purposes.

- **Payment of Bonus Act 1965**: The payment of Bonus Act provides for the payment of bonus to persons employed in certain establishments on the basis of profits or on the basis of production or productivity. The act is applicable to establishments employing 20 or more persons. The minimum bonus, which an employer is required to pay even if he suffers losses during the accounting year is 8.33% of the salary.

- **Payment of Gratuity Act 1972**: The Payment of Gratuity Act provides for a scheme for the payment of gratuity to all employees in all establishments employing ten or more employees to all types of workers. Gratuity is payable to an employee on his retirement/resignation at the rate of 15 days salary of the employee for each completed year of service subject to a maximum of Rs. 350,000.

- **Maternity Benefit Act 1961**: The Maternity Benefit Act regulates the employment of the women in certain establishments for a prescribed period before and after child birth and provides certain other benefits. The act does not apply to any factory or other establishment to which the Employees State Insurance Act 1948 is applicable. Every women employee who has actually worked in an establishment for a period of at least 80 days during the 12 months immediately preceding the date of her expected delivery, is entitled to receive maternity benefits under the act. The employer is thus required to pay maternity benefits and/or medical bonus and allow maternity leave and nursing breaks.

- **Aadhaar (Targeted Delivery of Financial and other Subsidies, benefits and services) Act, 2016**: "Aadhaar" is a Sanskrit derived word of Hindi language, which translated in to English language, means Foundation. In effect as it has been applied in India, Aadhaar is a 12-digit Unique Identity Number (UID Number) issued to almost all Indian residents based on their biometric (finger impressions scan of all ten fingers and iris scan) and demographic data. Each person enrolled under Aadhaar system is issued a Card known as 'AADHAAR CARD' which carries the 12-digit UID Number allotted to the person along with the card holder's picture. The 12-digit UID Number connects to a National Database wherein is stored the Aadhaar Card Holder's Biometric Data (finger print scan and iris scan), the latest Address and the Cellular (or Mobile) Phone Number. It is the world's largest biometric ID system, with over 1.19 billion enrolled members as of 30 November 2017, representing over 99% of Indians. It is the most sophisticated ID programme in the world and is considered a proof of residence and identity of its holder. However, the Aadhaar Card is not a proof of citizenship, and does not by itself grant any rights to domicile in India.

It is highly advisable for anyone who engages Indian workforce for any purpose in India to ensure that only the persons having proper Aadhaar Card are engaged or employed. Moreover, while so engaging personnel (supervisors as well as workmen) the employer must maintain a record of Aadhaar Card Number of each person (photocopying an Aadhaar Card is illegal). Doing so would ensure that whosoever is engaged or employed by some employer has an Indian Identity and his antecedents are verifiable and all those who are so engaged for any work are genuine persons.

It goes without saying, therefore, that under no circumstances any person, who does not possess an Aadhaar Card, should be employed for any work whatsoever. There are authorized agencies who can be entrusted to verify the Aadhaar Card of each person by cross tallying the actual biometrics with the recorded biometrics of the holder of any Aadhaar Card. Doing so will ensure that those with fake Aadhaar Card will be detected. Availing services of such agencies is highly recommended.

The above is not to be taken as a comprehensive list of Labour Laws, Rules and Acts. Such Laws, Rules and Acts are all subject to revisions and amendments from time to time and therefore it is necessary to check provisions of all latest amendments of each and every Act as also to check whether some of the above have become redundant and some newer Laws, Rules and Acts have become applicable.

In India, the Acts are promulgated by the Central Government and based upon them each State frames its own Rules which are to be adhered to and are to be satisfied whenever and wherever there takes place employment of workmen (females included).

It is highly advisable for whosoever employs workmen (females included), also has someone well acquainted with the Indian labour laws and who can make sure that all applicable laws are fully complied with at all times during continuation of the project.

As a final note about project contracting, if construction workers are from countries other than where a project is being constructed, they might be subject to the laws of their native country in addition to host-country laws. In the global arena, it is not always clear which legal system has the rights to settle disputes; therefore, international contracts usually specify that international arbitration or exclusive jurisdictions will be used to settle construction claims and disputes.

# PROJECT EXECUTION

The term 'Project' can be defined variously. For example, a project can be said to be an enterprise, individual or collaborative, that is carefully designed and planned to achieve a particular aim or result. A project can also be said to be a set of specially planned activities for creating some unique product or service. Project can still better be defined as sequence of cost and time bound tasks, planned from beginning to end within predefined resources and end results. Others may say that a project is a problem scheduled for solution (J. M. Juran).

As mentioned before, a project manager is basically required to manage the following:
- Breaking down the work (deliverables) required to be done for materialization of project into adequate plurality of discrete activities (work packages) to be performed and controlled, depending upon size and complexity of project. Level of hierarchical decomposition should be enough to assure the completeness of the deliverables (scope of project) and respective activities to be executed.
- Time frame estimate for carrying out of each activity. Planned durations will be defined based upon resources availability and respective work quantum.
- Sequence in which each activity is to be performed, thus some activities precede the others while some follow the others and some run simultaneously, often some idle time being provided during planning between the completion of a preceding activity and start of a

following activity, such idle time being called 'float'. Such logical sequence determination should take into account the mandatory or discretionary dependencies; external dependencies; as well as best practices (lessons learned) taken from similar projects.
- Planning of resources (manpower, machinery, material, implements, tools and tackles and money for them all). Such planned resources as well as respective durations for their availability should be allocated to each of the activities. As mentioned above, this will indicate the respective planned activity durations for each amount of works. As lessons learned, a Resource Breakdown Structure (RBS) should be created and indexed to all kind of resources by category, availability calendar and type, to be assigned to respective activities.

In order to be able to manage the above successfully, it is highly advisable to make use of some Project Management Software. Several reputed softwares are there in the market (e.g. Microsoft Project or Primavera) and different EPC contractors use one of them or the other as they please. Therefore, if project manager has to execute a project on behalf of an EPC contractor he/she would already be familiar with use of software for project management. If, however, project is to be executed in some other manner project manager may have to acquire one of the available software and put it to use for managing a project in hand. A lot of information is freely available through Internet and I expect that project manager would be able to make good use of it for computerization of his/her project.

In managing a project, using computer and software is but much insignificant a requirement and there is a lot more that a project manager has to take care of as an administrator. Main actions among these are the following:
- Project manager should ensure that all of the relevant project and team data (here 'data' means details of what-

ever is transpiring at project site) reaches him/her and he/she parses through it to provide concise and clear direction/(s) to each member of the team and advice or progress report to stakeholders, and management. Propagation of this information is up to the project manager and the organization. Project manager has to work really hard in order to be able to give precise response to any of the stakeholders, in case they want a quick and concise verbal reply about any specific question as to progress of project.

- Documentation is the foundation of all projects and provides the basis for project manager to plan activities of project. Documentation generally includes at least the following:
    - Correspondence with Owner and other relevant parties,
    - Copies of all Approvals of all kinds (including approval of drawings, diagrams and schemes) – be they from owner or owner's engineer or from authorities,
    - OEM's Manuals.

  Therefore, project manager must possess and review all the necessary documentation and make sure that he/she is in possession of all that is deemed 'required'. A project manager does not only manage or assume responsibility towards the content of documentation, but also that of its storage and ready availability. A good project manager makes sure that documentation is not sidelined in favor of delivery and if it does so happen it should be tracked lest project goes off track.

- Facilitation of Meetings. Project manager must make sure that all key team members who act as sub-managers of activities pertaining to various areas of project meet regularly and exchange their progress, problems and matters of concern. Presence of project manager in each and every such meeting is not necessary, rather he/she must assume responsibility for facilitating the meetings. Facilitation doesn't simply mean taking notes, but rather steering all such meeting by way of prepping their agenda, ensuring the right audience, driving the teams to value adding conversations, and a little bit of note taking at the end. When a team is not communicating effectively it may become necessary for the project manager to step in and force the communication. This may be in the form of daily standups or weekly status meetings. Project manager should force communication by way of putting the team into a meeting and properly facilitating so as to drive meetings to be short but concise and purposeful.

- Project manager must not simply drive his/her team for completion of tasks at all costs and in any manner whatsoever, but rather really lead the team forward for making sure that there occurs no neglect and all requirements of documentation are met without any smallest fail.
- Project manager also must maintain follow-ups on a regular basis. If there sets in slackness in follow-ups on part of project manager for any significant period of time, things generally go off the track and commitments are not met. The project manager should track these actions and make them part of the project plan. Most of these types of follow-ups are action items from meetings or discussions. These are not major milestones, but minor things that contribute to the success of a project and failing to complete these small items usually results in delays to a more major deliverable.
- In doing as above project manager would often come across situations where activity to be undertaken by one team member conflicts with an equally important activity to be undertaken by some other team member. A good project manager who ensures proper flow of communication more often than not comes to know of such conflicts in advance of their assuming a crit-

ical situation and he/she is able to proactively resolve such conflicts, often making use of time floats that may be available or through rescheduling several activities. Over a long enough project, a project manager is bound to find conflicts in the team and/or with management. The project manager should take an active role in reducing and resolving these conflicts.

- Even with best and most sincere efforts on part of project manager there do appear road blocks and issues. Project manager's success lies in being able to timely identify any broken, inefficient, or ineffective processes that may have arisen for any reason whatsoever. These items cause a problem and potential delays in projects. It may not always be able possible to correct the inefficiency but at the very least project manager should document and when possible he/she should drive for correction.
- Process Improvement. Project manager working as above is generally in a unique position to see processes that affect delivery of a project. Often process changes that seem benign but when applied at the project level, result in a significant impact to the project deliverables. As such, the project manager needs to advocate process improvements from this perspective.
- As the project moves through execution, a critical part of the job of project manager is to remove roadblocks and impediments for the team. This can come only through continually looking ahead of the team. Looking ahead will allow the project manager to remove these roadblocks before the team gets there. Basically, project manager has to ensure that team doesn't lose momentum because a roadblock if not removed well in time would lead to bringing the entire group of project teams to come to a halt and restart after removal of roadblock. This unnecessary wastage of energy must be avoided.
- Act as Marketing Department for Team and Project. A

good project manager acts as marketing department for his/her team and makes sure that project remains in forefront in eyes of management, customers, and stakeholders thus championing the successes of his/her project team and leading them to celebrations. Thus, a good project manager increases moral of his/her team and drives it to better teamwork and thus leads the team to greater outputs in the future.

Before proceeding further on the topic, it needs to be fully understood and appreciated that foundation of all projects is the contract for execution of project since end result, cost and applications limitations are defined in it. Therefore, project manager must have the concerned contract at his fingertips and should be careful all the time to make sure that all provisions of contract are being strictly met.

When it comes to the execution to the project there are several ways of succeeding in terms of developing and completing deliverables. The execution phase of project involves coordinating people and resources, as well as integrating and performing the activities of the project in accordance with the project plan. The key thought to keep in mind here is that you will be on track towards a successful completion, as long as your team works effectively and adheres to the plan, and this is an attribute to the project managers. In order to present lessons learned during the implementation of projects, the authors have chosen projects in India, which in several aspects replicate same conditions as in other developing countries, where most challenges can be found in a similar manner. Its peculiarities may enlighten and provide awareness to the readers when facing similar situations.

## Handling Manpower (Lessons from India)

The very first requisite for successful execution of any project is that the right kind of manpower is engaged at different levels for its execution. This apparently doesn't seem to be a problem due to India's population and corresponding easy availability of bulk of personnel with various qualifications. However, things are not as simple as they appear and lack of understanding of the peculiar nature of manpower that is available in India can result in disaster for a project, even though Indians as such are quite smart, intelligent, diligent, honest and reliable. This aspect is particularly relevant to foreign companies developing projects in India, who due to their ignorance of local work culture diversity (habits and beliefs), can get trapped in serious miscommunications, misinterpretations, frustration and needless disputes.

Although globalization is clearly contributing to increased integration of labor market and reducing the asymmetries between developed and developing countries, what is still at stake are the cultural differences which have to be respected, understood and the respective management procedures have to be adapted to better deal with this reality. Appendix 1 is dedicated to this aspect of the matter.

The very first requirement is that the personnel who are supposed to undertake and complete various tasks are 'competent'. Generally, we judge competence by a person's level of education, his/her earlier employment history, references and recommendations, performance at the interview, general appearance and impression given by the person as a prerequisite for offering employment. No doubt, the above attributes are important and go a long way in selection of the right personnel for various tasks of various levels.

Appendix-2 gives brief extract form Indian scriptures defining three basic tendencies of humans that are present in each of us in four combinations so that we have evolved with four basic types – for convenience classified as Alpha, Beta, Gamma and Delta types. My purpose of mentioning, hereinabove, about humans having evolved with four basic types with a mix of three basic tendencies is the following. Irrespective of the country we consider, the fact remains that about 90% of the people are of Type Delta. This is what results in what is termed as Pyramidal Management. In all organizations working on older system of management people are supposed to give their best output, in contrast the organizations that have adopted the modern system of scientific management adopting ways and means to actively assist people to give their best using modern management tools and techniques that produce an atmosphere of hand-holding thus enabling people at lower rungs of management feel constantly guided and supported thus being able to result in exceptional output. Such management systems do not work in a Pyramidal manner and have a relative horizontal organisation with fewer rungs of managerial levels.

In first-world countries where population is within reasonable limits with low unemployment and high minimum wages, people tend to study only that which they genuinely like and are capable of excelling. In these situations, the educational qualification of a person is, in general (although never always), good enough an indicator of real capability of the person. Therefore, the employer is sure about what he would get from the unskilled, semi-skilled, skilled and highly skilled and specialist employees and what he would get from graduates and post-graduates in different fields of study. In these countries persons who, in their younger age could not get educated beyond Middle or High School, find it convenient to educate themselves through remote learning systems to achieve higher education as they age. As a result, many of them succeed to become supervisors in due course and some even achieve the position of being experts of their trade. It is through such persons that the management is able to provide an atmosphere of hand-holding to younger workers.

In India, unfortunately, there is an uncontrolled population explosion with population of the country having already reached a situation where it is unsustainable. As a result, plurality of youth seeking education far exceed the plurality of educational institutes of excellence. As a result, education in India has become an industry, without any quality control and with sole aim of high profits. A large number of institutes of all kinds have mushroomed all over India unscrupulously, in connivance with politicians. These institutes include a large number of High Schools and Colleges, Professional Institutes such as Engineering and Medical Colleges and Universities that churn out youth with qualifications starting from High School pass-out to Post Graduates and even Doctoral qualifications in various fields of studies, all with pitiable knowledge. The youth who qualify from such institutions may or may-not be suitable for handling the job expected of them based on their qualifications.

There, however, are institutes of excellence and high level of learning under Central Government but their plurality is much too low and seats available in these premier institutes are too few to cater for the huge crowd of Indian youth and it needs a certain high degree of bare minimum capability for under graduates to get admission in these premier institutes, which enables only brilliant youth to succeed to get admitted in these institutes. Nonetheless, the graduates, postgraduates and research scholars produced by these institutions are comparable with the best in the world and many of them leave India for greener pastures abroad.

Good news is that what has been described above is the situation only with personnel needed to handle managerial jobs. For recruiting workforce comprising semi-skilled, skilled and highly skilled mechanics, electricians, welders, carpenters and other tradesmen the good news is that there are training institutes known as Industrial Training Institutes (ITIs) in good enough plurality and fortunately the youth who are in unnecessary rat race for becoming white collar managers just do not seek admission in these ITIs, therefore, education industry producing ITIs certificate holders in bulk and without quality control, have not come up. Hence, the young men and women possessing diploma or certificate from ITIs are generally good and proficient in their respective trade. Premier among such construction skills training institutes are several L&T Construction Skills Training Institute in Ahmedabad, Gujarat.

With the knowledge of state of affairs as described above it is not difficult for a project manager, who has chosen to undertake execution of a project in India, to filter out candidates based of some database of institutions.

Last but not least, it is prudent to be aware of and acknowledge the basic human personality types, traits and tendencies so as to judge whether a person chosen for a certain job has the capability of handling the job successfully and reliably. There are many personality classification systems like Keirsey Temperament Sorter, the Myers-Briggs Type Indicator, and India's ancient system of 'Gunas', which is described in Appendix 2. It is advisable that the project manager keeps a keen eye, in making a judgment about each person selected, for ensuring that the candidate has the basic making for the type of job he/she has been chosen. This is particularly important in respect of persons chosen for supervisory and managerial jobs. This has to be ascertained over a certain, though small, period of time, say a fortnight or at the most a month, by which time the behaviour and reactions of the person under various realistic circumstances comes out in open.

Project manager would do good by bringing Experts who have gained high level of proficiency in their trade, to provide an atmosphere of hand-holding to Indian workmen engaged for skilled work in a project.

Despite having done all the above there may still remain, the problem of personal perception of workmen (and even supervisors) about what is right or adequate in relation with various jobs or activities being performed. I would like to describe my real-life experience in order to explain this.

*In a power project a large capacity power transformer was under installation. All related work had been completed at evening just past end of working time for the day. Just then the supervisor who was entrusted with this job noticed that insulating oil was leaking from a flange of oil forced cooling system, the flange being on the pipe welded to upper part of transformer tank for the purpose of fixing a gate valve through with insulating oil was piped to oil cooling radiator outlet of which was similarly piped to lower side of the transformer tank via a flanged gate valve. Extent of oil leakage, in opinion of the supervisor was quite "low" and he decided to call it a day with the resolve that the problem of oil leakage would be taken up as the first thing next morning.*

*Fortunately, the supervisor thought of bringing the matter to my notice and came to inform me about the leakage and his action plan for attending to it, however, while narrating the matter to me he told that oil leakage was 'very minor'. I asked him about what was his criterion to consider a leakage of transformer oil as 'very minor' to which his reply was that the leakage is 'very little'. I then enquired of him the criteria for considering the leakage as 'very little'. This confused him and I observed signs of annoyance on his face perhaps because in his opinion I was questioning him for no real reason as 'minor' or 'little' meant same and how else did I want the matter to be reported and what why I was smarting him by asking such absurd question as 'criterion' for a leakage being 'very small' or 'very small'. I sensed about his annoyance and to make his life easy requested him to tell me the 'rate of leakage' in terms of volume of oil that would leak in one hour. This he had never expected and told me that there was no reason to worry about, as the leakage was really 'very little'.*

*I sensed that the person had throughout his life experienced only handling things based on 'rules of thumb' rather than having ever seen any 'scientific approach'. Therefore, I decided to go to the site of this particular work along with the said supervisor. I found that about 25 drops of oil were leaking per minute, which considering about 20 drops per ml of volume indicated that about 1.5 liters of oil would drop in 20 hours (assuming that next working hours would start after a gap of 16 hours and a time of about 4 hours would further pass before oil level in transformer is lowered below the flange level. I considered this leakage to be intolerable because without knowing cause of the leakage it was not wise to blindly assume that leakage rate would not increase further overnight. Therefore, I decided to take immediate steps to get oil drained from transformer oil tank and pump-radiator assembly such that oil level was brought down below level of the leaky flange, with programme to proceed with corrective steps next morning.*

*When work on corrective action was started next morning, it was found that due to neglect of someone involved in installation of the transformer, a twig of about 2 mm diameter (perhaps from grass) had somehow entered between the neoprene gasket and flange on transformer tank side. The gasket was of 10 mm thickness, held in circular recesses on mating faces of both flanges, depth of each recess being 4 mm so that a pinch of 2 mm was occurring over the gasket. Obviously, the transformer oil had taken a couple of hours in traversing full width of the Neoprene Gasket thus leading to inference that if it took about 2 hours to increase rate of oil leakage from Zero to 20 drops per minute, the rate of leakage could as well double over 20 hours.*

The above incident has been described to emphasize the point that common tendency in India even for well experienced supervisors is work by rule of thumb principles rather than on the basis of scientific analysis and project manager has to bear this fact in mind since habits die hard.

As final note regarding labor handling, considering that labor is the single largest component on the overall costs of construction of a project. For a grassroots new industrial project, for example – a thermal power project - the labor costs can range anywhere from 35% to 65% of the construction costs, depending on where the job is located, where the labor comes from and the type of plant in construction: for a combined cycle job, the labor would be on the lower end of the scale, whereas a new coal-fired project would be more labor intensive. Therefore, when the labor is not properly managed, mostly due to poor supervision and leadership, the remaining other segments will be badly impacted along with the project itself.

## Sourcing goods & machinery from developing countries

Industrial Projects are commercial ventures and therefore it is very important that those handling any Industrial Project are able to properly workout the returns to decide to make in-house or buy items externally. If a project is being executed in house, by owner it is necessary to divide the estimates of costs broadly in two parts namely (i) capital expenses towards cost of all plant and machinery and (ii) expenses towards cost of execution.

If plant and machinery is imported proper selection of technology and plant and machinery is the key to ensuring best results and then the grey area that remains is execution of the project. In general, the manufacturers and suppliers of plant and machinery based in Europe, USA, Canada and Japan maintain their own good control of quality not only as a part of maintaining their reputation in the market but also with the point of view on meeting of guaranteed performance and the warranties in order to avoid penalties and arbitrations arising out of disputes due to lack of performance and fulfillment of warranties. However, with Korea and China the case may, as an exception, be somewhat different and more vigilance may be required.

In contrast if the plant and machinery is of Indian make - more or less same being applicable to other developing countries - then there can be unpredictability in many matters related to consistency of quality, meeting of guaranteed performance and even timely delivery. Shirking responsibility by manufacturer in fulfillment of Warranties. Therefore, irrespective of the manner in which a project is executed the agency responsible for execution will do good by factoring the element of unpredictability in such cases.

In my experience it is best practice, in case plant and machinery is sourced from India, to search those manufacturers who already have supplied identical or exactly the same machinery to buyers in European Countries and/or USA. However, it would be necessary in such cases to carry out the due diligence to really ascertain that such supply abroad has met the buyer's appreciation rather than complaints and blacklisting. It would be best to select such suppliers who have record of having received repeat orders, recently rather than too long in the past. It is further advisable to enter into correspondence with the buyer/buyers so as to hear the truth from horse's mouth.

A project being executed in India will be an activity wherein involvement of humans, as limited by the current state of the art and technology, would be unavoidable. Therefore, those of the readers who have never earlier experienced handing any serious project in India involving sufficient interaction with Indian employees of all types and levels (status, education, profession, experience, gender and religion), may not be having any idea of the cultural, physical and psychological behavior, habits, and tendencies of Indian employees as compared, particularly with the Western world. There is so much of difference from what happens in west that the first reaction can be that of a cultural shock.

One human behavioral aspect of Indians that is very different from people of the western world is UNPREDICTABLY which makes doing any business in India difficult because with unpredictability prevailing in India it becomes near impossible to calculate the return on investment. Notwithstanding structural uncertainties, for which the possibility of the event is based on cause / effect chain of reasoning, one should put attention on risks where there is enough historical precedence, in the form of similar events and consequences, to enable one to estimate probabilities (even if only judgmentally) for various possible outcomes, to be considered along the calculation phase as risk mitigation framework. One should note that this

UNPREDICTABILITY is not a prerogative of Indian society only, but it is also present in most other developing countries.
I have tried to explain the matter threadbare as below:

- If the problem is only relatively poorer infrastructure, it can be factored into shipping cost & delivery time.
- The problem with lack of adequately experienced and trained labors having proficiency in their respective job can be factored into capital/startup cost.
- However strange it may seem to project managers of foreign companies handing projects in India, the problem of Corruption too can be considered not to be a real problem provided it is possible to predict and tell who is needed to be paid and how much as that can be provided for to become part of operating cost. From my friends in other countries I understand that such practices, although not uncommon in all countries to some extent, are pervasive in most developing countries, as a result of widespread poverty, unresponsiveness of governments to the public and the absence of effective and responsive administrative systems. In this regard, and based upon strict compliance of rules established after past scandalous bribery developments occurred in some countries, such practice nowadays is imperative for some European companies to make business under these circumstances. In addition, bribery and corruption are to some extent indigenous within certain political and social settings.
- The problem of lots of burden of paperwork, due to red tape, too can be smoothly handled and its cost can be factored as the longer startup time the lower the return on investment and that can be reasonably correctly calculated, and can be made up with higher profit margin.
- However, the real problem in execution of a project in India is that almost NOTHING CAN BE CALCULATED.
- When someone you are dealing in connection with a

project in India says "yes", it is possible that the person himself/herself may not be clear about whether he/she really meant "yes" or "no". This sort of miscommunication is related with most of failures and disturbances within Indian, and not only, projects run by Western companies.
- When a worker says that "I'll be there in 5 minutes", you have no idea whether he/she will keep his/her word, or he/she will show up, may be, some hours later. It is similar as the famous 'mañana' which in Spanish spoken countries can mean anytime, but not now.
- When an official approaches you and points out certain deficiencies in something pertaining to the project and in your best judgment you find that the deficiency being so pointed out is of no real consequence and in similar projects in other countries you had followed exactly the same practices and no one had ever objected and you consult your Indian administrative employee (or handler) and you are told that sorting out the matter would need some expense in the form of unofficial "expedition fee", you have no idea how much he's asking for, and how much faster he can expedite, and how many more instances of this type may be in the offing.
- When you are presented with a pile of paperwork, you have no idea whether this is the last pile, or whether there may be, umpteen more of them that will follow sooner or later, and whether you were even given the correct paper works to fill out, in the first place.
- Even if you have someone coming in to help, you have no idea whether he'll hit the nail right or just way here and there and then forced to wait for another, hours or even weeks, for someone to show up and complete the ritual.
- All that plus the occasional "I refuse to work with people from that caste" plus workers union striking the work, construction power outages, construction water supply stoppages make doing business in India ex-

tremely unpredictable.

If you are a business owner, would you buy from someone who sells you the same item with $5, knowing that the goods will likely be delivered on time and to specification, or are you going to buy from someone who sells you for $4, but you have no idea whether you'll get that item and when and in what condition of readiness for being put to intended use as soon as delivered? Most people would buy it for $5, then just markup the profit. Business is all about predictability, trying to make money without ability to predict is called gamble, and business owner don't like to gamble when they could avoid it.

However, the above knowledge doesn't help anyone since it doesn't provide any clue about how to proceed if sourcing from India is the fundamental principle or requirement of executing a certain project.

The situation is not as formidable as it looks, provided you first make your mind free of prejudices and allow your brain to use intelligence for successful handling the situation to ultimate advantage and good profitability. However, some special efforts would be required to ensure success. Here again the cultural diversity, as well as any other factor (physical, economic, institutional and political) that could influence the environment of a particular project, presents a complex challenge for project managers. Lack of understanding can induce conflict arising from misunderstandings and differing priorities, as well as serious miscommunication, misinterpretations, frustration and disputes, which can obstruct the growth and the productivity of an organization. These are explained below.

## *Due diligence in Selection of Manufacturer/Supplier.*

Due diligence is called for making sure that the party has supplied good enough plurality of identical or better exactly same product to others, preferably abroad.

The next important step is to obtain contact details and addresses of earlier clients and then to obtain feedback from several such previous clients of the party, no lesser than three. After receipt of feedback if majority of previous clients give a positive feedback still it is advisable to arrange a visit to the site of installation of the product at premises of such clients as have given positive feedback so as to make sure that the feedback obtained are truthful and the product is successfully functioning and giving services as are intended to be obtained for the subject product.

However, should a situation arise where general feedback about the product is good but the product being used by others is not exactly the same as is intended to be procured for the subject product, it is necessary to frankly discuss the matter with the concerned supplier/manufacturer and to find out about the steps that the party would take to make sure that the product that would be produced by it for the first time would be right and flawless. Even if the explanation given by the party is logical and technically correct business sense calls for coverage of the risk since in a freak chance despite everything being logically right the item or product produced for the first time, to the exact specifications as called for the subject project, may not be right and may be having some unforeseen defect.

## *Step-by-step procedure suggested for selection of manufacturer/ supplier.*

### Adjudge the capability of manufacturer/ supplier

The very first step in selection is a detailed list of what to look for in order to adjudge capability of any manufacturer/ supplier. It should include full knowledge of
- requirement of raw materials and their bare minimum quality parameters,
- names of acceptable suppliers of respective materials,
- quality assurance requirements for acceptance of raw

materials,
- storage requirements for ensuring deterioration free storage until respective materials are put to use for manufacture,
- manufacturing process that should be considered as acceptable,
- quality control steps during various stages of manufacture/ production,
- documentation requirements in connection with all the previous steps,
- post production storage requirements,
- packing and transporting requirements,
- special requirements pertaining to transport vehicle, if any.

## Inspection of works of manufacturer/ supplier

In your establishment, there should be person/ persons familiar enough with the above, having experience of inspecting manufacturing facility of prospective manufacturers so that there do not remain any aspects hidden from his/her prying eyes (although this expression means offensively curious or inquisitive, it has been intentionally used here to impress upon necessity of inspection, in depth rather than on the surface).

You should expect the possibility that your inspector will be given welcome and all treatment befitting to a king including best quality food and drinks being served in most appropriate manner taking special care that ego of the inspector is sufficiently satisfied and he/she slowly but surely develops a humane tendency of being clement or lenient in his/her views regarding the party. Understanding this possibility your inspector should be a person of rather matter-of-fact bent of mind and his/ her focus on the task assigned to him/ her should never be lost.

It is advisable that time of carrying out of inspection is fixed to be early hours of the day so that the inspector can, may be after some bare minimum formal introduction, ask the party to enable him/ her to inspect the facility, see things, ask questions, take a look at documentation and records (with particular care about their dates) and procedure. About this the inspector should devote maximum time before any formalities of refreshments that can preferably be postponed after inspection.

The inspector should carry a personal notebook on which he/she should note down various observations and questions to be asked. When physical inspection is over, time permitting, the inspector should request meeting for discussions and clarifications and the clarifications furnished should be noted down clearly so that full knowledge gained during inspection and discussions remains firm rather than subjective due to dilutions on memory that do occur over time.

The notes taken would come handy when premises of two or more parties have been inspected, for tabulating the observations with a view to compare between the parties for the purpose of selection of manufacturer/ supplier.

Mentioned above are the bare minimum recommendations for ensuring effective inspection. More may be added from own experience of the project manager and the project team.

**Obtaining detailed schedule of work from selected manufacturer/ supplier.**

Fully detailed schedule of work, including all steps starting from
- design and engineering (if involved),
- preparation of bills of material,
- obtaining quotes from prospective suppliers of raw materials and other ingredients and goods needed for pro-

duction of the item ordered,
- delivery schedule of such materials, ingredients and goods,
- schedule of each step of manufacture,
- schedule of quality assurance inspections/ tests and its documentation,
- inspection by your inspector during various important stages of production,
- final inspection and acceptance tests in presence of your inspector, should be obtained from the party and mutually agreed upon, if necessary by additions of any tests and/or inspection that in your opinion might have been omitted.

**Special provisions in the contract/ purchase order.**

Once the manufacturer/ supplier has been properly chosen, it is advisable that apart from the customary provisions, the following provisions are also included in the contract so as to cover the risk of buyer. Such coverage of risk is not likely to come for free and the manufacturer/ supplier is likely to initially just disagree to accept these provisions and then ultimately come up with the plea that such risks are covered through insurance by the buyer at its own cost and not by manufacturer/ supplier who is not in the business of providing insurance. However, good negotiating skills may lead to the manufacturer/ supplier ultimately agree to these provisions, may be against a small increase in the price – which the buyer should weigh in terms of gain vs. loss and arrive at its own judgment about omitting or including these provisions.

- Release of all payments only against irrevocable Letter of Credit (LC) from a Scheduled Bank of the country concerned, to your approval. The LC should have validity till a fortnight after the mutually agreed period of warranty. This would not only safeguard your money but will also keep the manufacturer/ supplier on his toes, thereby ensuring that the party dare not slacken at any step and does not compromise of quality or delivery period.

- Provision for posting of your own reporter at the premises of the manufacturer/ supplier, as further detailed below.

A reporter so posted at premises of the manufacturer/ supplier need himself/ herself not to be an expert of the process. Rather he/she should be a smart and intelligent person who should be capable of inspecting and comparing various activities vis-à-vis the agreed Schedule of Work of the party and capable of properly reporting all that goes on at the works of manufacturer/ supplier in relation with the product ordered. For this purpose a structured reporting format should be prepared and the reporter should send reports in accordance with it at least twice daily preferable one around midday and the other around two hours before closing hours of work at the project site.

While on the topic of placing a person at the facility of manufacturer/ supplier, I would like to describe an instance from my own experience, as below.

> *At that time, I was working for an Aluminum Industry engaged in manufacturer of Aluminum metal from its ore Bauxite. The process in the last stage is electrolytic and requires supply of electricity in bulk, for which the industry had its captive power plant for supplying electricity to the smelters. The Industry decided to add some smelters and also matching electricity generation capacity in its captive power plant. I was entrusted with the responsibility of adding a pulverized coal fired electricity generating unit to the captive power plant. Date of commissioning and putting into commercial of the said power generating unit had to coincide with the date of readiness of the new smelter lines so that the investment becomes productive from day one. The entire power plant machinery was sourced indigenously.*

87

*Luckily, we had posted a person at boiler manufacturer's works to report progress regarding each step and to function as my eyes and ears. At one point of time when various components of super-heater were under production he reported that suddenly the job of robotic welding of nipples in super heater headers had come to a standstill and he was unable to find out the cause. He also confirmed that there was no labour unrest and all was peaceful. The report was alarming, therefore I made a phone call to general manager (GM) of the boiler manufacturing facility and expressed my concern about the report received by me. The GM told me that an electronic controller card had failed and spare card was not available in the facility warehouse and that steps were being taken to import a spare from the OEM in Japan. The boiler manufacturing facility was Government owned due to which the process of importing the spare card was tedious and time taking. I sensed a delay of couple of months and discussed the matter with my boss and suggested that it would be in our interest to procure the said spare electronic card from Japan and to gift it free of cost to the boiler manufacturing facility at no cost to it. My boss readily agreed to my suggestion and allowed me to proceed. I talked to the GM of the boiler manufacturing facility and took appropriate action to ensure that the spare card reached the GM in three days. This resulted in prompt restart of the work at the boiler manufacturing facility and the GM was so kind as to order two shift working in order to mitigate the lost time. Thus, rather than there being a long delay the super heater coils reached by site a couple of days in advance of my internal schedule.*

- Provision for cancellation of contract and right of purchaser to procure the same goods from another party of purchaser's choice at total cost of the manufacturer/ supplier, should the purchaser find that there are unacceptable quality lapses or chances of delay in successful production/ manufacture and delivery of the goods on order. Of course, it goes without saying that

the manufacturer/ supplier would not agree to such terms unless there also are provisions of adequate notices from the purchaser to the manufacturer/ supplier intimating onset of provisions cancellation of contract and right of purchaser to procure the same goods from another party of purchaser's choice at total cost of the manufacturer/ supplier. However, such a provision in the contract would go a long way in keeping the manufacturer/ supplier on its toes and free from complacence.
- Provision of safeguarding interests of the buyer against latent defects that may though be present in the product but do not show up even up to warranty period. However, a proper time limit will have to be fixed and mutually agreed to between the parties for coming into light of latent defect/(s) and liability of the manufacturer/ supplier to provide replacement of the product at no cost to the buyer.
- Safeguarding against any important electronic black-box component, bought out by the manufacturer/ supplier, going obsolete or out of production in market.

The first step is to make a provision in the contract to the effect that the manufacturer/ supplier shall not use any black-box components that are totally proprietary in nature and that only such components shall be used for which there exist no lesser than three manufacturers.

This would need own due diligence in terms of finding out (i) vintage of various black-box components involved (trend in electronics is that it hardly takes more than five years for anything electronic to go obsolete), (ii) financial health of their respective manufacturers, (iii) availability of exact replacement of black-box components from other manufacturers of repute and (iv) possibility of replacing black-box components with several separate discrete components.

- Safeguarding against manufacturer/ supplier going bankrupt and being unable to provide after sales services. Truthfully this cannot be achieved by any extent of clever wordings or additional protective contractual clauses due to legal shield provided by laws to a party that goes bankrupt for any reason.

The only remedy can be making sure before placing the order or entering into contract that there is nothing within the scope of works and supplies of the manufacturer/ supplier that is so proprietary that it may be impossible to get any services that one can expect in terms of after sales services from the manufacturer/ supplier. If there do exist such works and supplies that would not be available from any third party in case of unfortunate insolvency of the manufacturer/ supplier, it is advisable to steer clear of such party unless some ways and means are identified to come handy in such a situation.

Experience dealing with projects in developing countries have shown the need to enforce contractually with site service providers some disciplinary aspects related with site management issues like health safety & environmental requirements among other site regulations. Also important is to clearly define and document the performance expected (metrics and accountabilities) when entering into agreements for the purchase of goods and services to be used at the site.

Roles and responsibilities need to be addressed as explicitly as possible to avoid reopening negotiations during the heat of the project. As an example, if the order is to supply of radiographic services, the agreement should spell out exactly how the service is to be performed, in accordance with specified codes or other criteria, and the skill level of the personnel performing the work. Additionally, response time needs to be established such as: "personnel to be on-site within four hours of initial telephonic notification, responsibility for barricades must be clear, and the time for delivering the film or digital data interpretation should be such that production work is not delayed.

Although most project procedures require the typical vendor selection process to follow a "three quotes and select the lowest bidder" scenario, that is exactly what often drives the relationship to be adversarial. An alternative approach is to use the bidding process only for identifying and prequalifying the suppliers. Then the next step would be geared to maximizing value creation, as opposed to reducing costs by means of squeezing supplier margins and scope. For this mutually rewarding relationship to happen, it requires open communication and transparency during negotiation and execution phases of projects in order to increase the likelihood of developing a strategic supply-chain relationship, as opposed to a one-off purchase with limited long-term value. Once supplier is perceived as a "key supplier", those vendors are gained the loyalty of their purchasers and a long-term business relationship may occur.

## Execution of actions for meeting contractual requirements

In effect execution of a project comprises successful execution of various Deliverables (Project Deliverables and Process Deliverables) by the parties, who are contractually responsible for respective Deliverables. Step-by-step procedure that may be followed for ensuring it can be as below.

(a) The first step therefore should be comprehensively listing down all the deliverables (work packages), with a meaningful level of decomposition in respect of its respective activities, and name of the respective party that is supposed to be contractually responsible for their execution; making sure that even the most minor deliverable doesn't skip inclusion in the list.

(b) Once all the deliverables have been listed, the next step advisable is to carry out brainstorming to establish predecessor-successor relationship between various deliverables. One has to remember that scarcity of qualified personnel and poor availability of resources in developing countries is very expected.

(c) Upon completion of comprehensive list of all project deliverables (including names of party that has to deliver each of them) the their predecessor-successor relationships, the next step advisable is to chalk out a schedule of execution of each deliverable, starting from the very first minor activity to the last minor activity say, final shake hands and celebration (of course inclusion of such minor details is certainly not essential and purpose of mentioning them here is to emphasize that no unavoidable deliverable should get excluded even inadvertently at this stage). When this moment is achieved, a clear preliminary critical path is defined which should be updated regularly anytime a deviation occurs on the planned activities in order to highlight and focus on the critical activities his attention should be allocated.

With this step, the project takes a shape, although an imperceptible one, and a very important though invisible and abstract foundation is laid, which will go a long way in ensuring successful execution of the project.

(d) The above mentioned three steps take the project manager to a stage when the time is ripe for calling a meeting of concerned personnel of all the parties for a threadbare discussion and dissection of the list of deliverables showing their respective desirable execution schedules and names of parties who are supposed to be responsible for properly putting in place each deliverable well in time and in the required schedule.

The meeting and discussions should lead to consensus between all respective parties about (i) that there remained no deliverable yet to be included for successful completion of the project, (ii) scope of each party towards the project so far as deliverables are concerned, (iii) schedule of putting in place of each deliverable for ensuring that nothing remains in the jigsaw puzzle of the project execution that has not fallen in place.

One should note that as far as developing countries are concerned, all sort of constraints exist to disturb the normal flow of project management processes. The best way to mitigate them is to strongly emphasize the planning phase by means of usage the best capabilities and experiences already collected in similar projects. Although it seems obvious that planning is key, one should note that the plan itself is nothing, as it will be changed inexorably along the ongoing project, but the act of planning ceaselessly is what makes the difference.

## Taking over project site to start project construction work

Monsoon season in most parts of India causes heavy downpour that often continues not only for hours but also for days. Sometimes a phenomenon known as cloud-burst occurs and more than 60% - 80% of rainfall expected for whole of the monsoon season takes place in just a few hours, that more or less interrupts most of the facilities such as drainage, power supply, vehicular transport in the city concerned or highway concerned and even railway traffic by flood waters covering railway lines. Disturbance to air traffic too occurs resulting in diversion of incoming flights to nearby airports and delaying departing flights by hours.

Therefore, in case the project has to be executed on a virgin site, it would be best if the site is taken over for starting construction work just after end of rainy season since it generally is not possible to start any initial civil works if site is taken just at or near start of monsoon season. However, it is highly advisable to visit a virgin project site several times during peak of monsoon season irrespective of the inconveniences that may be faced. The experiences gained from such visits may go a long way in making execution of civil works a real success. Such visits are likely to help making it clear as to extent of flooding that would need to be totally eliminated when the next monsoon season arrives and the site sees peak of the downpour of monsoon season. These visits may also be helpful by way of enabling a feel as to storm water floods mitigation measures that would be needed at work site.

The pre-contract meeting (Kick Off) is an important meeting that takes place after the contractor has been appointed but before work commences on site. It is an opportunity for the project team to meet (perhaps for the first time) and to plan the construction stage.

The pre-contract meeting is chaired by the project manager and is an opportunity to:
- Make introductions and issue contact details (perhaps a project directory).
- Clarify roles, responsibilities and lines of communication.
- Agree meeting schedules, meeting structures and attendees.
- Hand over outstanding documents (such as insurance certificates and bonds) and issue outstanding information (which may be including any variations made since the contract was awarded).
- Issue nomination instructions.
- Discuss the contractor's master programme, including incorporation of works outside of the main contract, inspections, commissioning and testing.
- Discuss the role of the project team members (including site inspectors).
- Agree procedures for monitoring, issuing, receiving and reviewing information (including the information release schedule if there is one, and its relationship with the contractor's master programme). This may include a distribution matrix.
- Agree site access procedures, regulations, respective lay down areas, site power and issues.
- Agree site induction procedures and other environmental, health and safety issues (housekeeping, traffic flow, waste disposal, etc.).
- Agree site security measures to be adhered by contractors / suppliers / service providers.
- Agree procedures for dealing with queries.
- Agree procedures for issuing instructions.
- Agree procedures for changing orders / extra work.
- Agree procedure for differing site conditions.
- Agree procedure for construction sequencing / project

access.
- Agree procedure for construction defects.
- Agree procedure for subcontractor substitution.
- Hand over contractor's procurement schedule.
- Mobilisation schedule and status.

The meeting should be minuted so that there is a clear record of the procedures agreed and decisions made. These minutes may form part of the contract documents, subject to agreement by both parties. At the time of taking over the parties concerned must be fully clear about their respective total responsibility in respect of the following:
- Security and safety of project site and responsibility in respect of incidence of any theft or any other criminal occurrence,
- Sanitation and cleaning of project site,
- Providing first aid and ambulance facilities for the project,
- Providing fire alarm and fighting machinery for the project,
- Providing facilities such as canteen, shrines (should be avoided, if possible, in highly diverse country like India as requirement of shrines can be large) and crèche facility for taking care of children of female laborers, if engaged at site for project work,
- Providing first aid facilities at project including a certified medical practitioner,
- Providing adequate water having the right properties and adequate reliable electrical power for carrying out necessary civil and other construction work,
- Required permits and other statutory requirements

## Civil construction work - Preliminary and preparatory work

The first step in civil construction work, after taking over and having all surveying reference points, based on national coordinate system, as well as information about existing underground structures (alternatively, carrying out explorative excavations), comprises earth digging, transport, filling, compaction, and surfacing of various areas in the manner depending upon use, to which each temporary or permanent particular site area is supposed to be put. Usage of various areas and their preparation can be listed somewhat as below.
- Lay-down Area for long term storage of materials and machinery: These have to be hard enough to bear dead loads of goods over long enough periods and may also need some graded paving to raise them above surrounding areas so as to ensure proper draining out of storm water, in order to assure 'high and dry' conditions for stored items.
- Temporary driveways and fences: These have to be good enough to be used for movement of heavy loaded trucks as well as construction machinery and cranes. Obviously, levels of such driveways should be such that even in worst rainfall there is no long-term water logging. Actually, the priorities to provide good accessibilities at sites in countries like India constitute a big pitfall that Western companies usually do not take into utmost consideration, generating unjustifiable delays. It is important that the temporary routes are planned carefully, including any intended change due to ongoing project progress, keeping in mind that they should meet the same basic safety standards applicable to final routes, namely: have firm and even surfaces, be properly drained, and have no slopes that are too steep. And, of course, be maintained along its usage.

- Temporary storm water drains and sewage systems: To the extent possible, it is advisable to locate temporary storm water drains to coincide with the finally planned permanent drains and construction of permanent drains should be prioritized. It would go a long way in ensuring smooth working at project site if all permanent storm water drains and culverts are ready duly constructed before arrival of monsoon.

- Digging for laying underground pipes and construction of electrical cable trenches, tunnels, if required, in engineering and design, however, I consider it desirable to lay all electrical cables in overhead cable trays suitably supported over cable racks, cable cross-over bridges and cable under-pass tunnels. Cable trenches and cable under-pass tunnels can prove to be highways to hell, if not engineered properly - making sure that even under the worst possible storm and rainfall there would be no flooding in them. Sometimes trenches are engineered taking it for granted that during incident of rainfall and storm negligible quantity of storm water would enter them because by virtue of being duly protected by trench covers. In real life, however, entire length of cable trenches remain without any trench covers during project execution stage when cables are being laid. Properly engineered cable trenches are provided with a gentle slope (say 1 in 5000) towards one end where all cable trenches are made to join a single deep and wide enough trench purpose of which is to act not as cable trench but as storm water drain leading to a catch pit of good enough dimensions, deep enough and so engineered that in case pumping arrangement provided in it works properly all cable trench would be free of flooding irrespective of extent of storm water rain fall. The catch pit has RCC (Reinforce Cement Concrete) walls with removable RCC roof, if necessary in the form of one or more roof slabs and houses three submersible storm water pumps each of a capacity such that a single pump is able to pump out storm water considering collecting in the sump during the worst possible storm and equipped with level switches to ensure adequate submergence of pump before getting started and tripping of pump upon water level having dropped to desired submergence level, a third level switch having been provided to start the second pump when water level rises despite one pump already being in operation either as a result of extent of incoming storm water being unforeseeably excessive or any other reason, whatsoever. Purpose of the third pump is to act as a common standby in the event of failure of one of the earlier said two pumps, the third pump too being duly connected to power system and equipped with level switch and in a state of readiness to function whenever needed.

- However, in certain areas, for example outdoor high voltage substations and switchyards, construction of cable trenches is generally unavoidable. Biggest problem in civil construction of such substations and switchyards is that no work in connection with construction of cable trenches can be undertaken unless all digging for civil foundations of structures of the substation and switchyard, construction of civil foundations, backfilling and consolidation of backfilled soil have been completed – the situation getting worsened if piles are needed for civil foundations of structures. There are several methods of carrying out the necessary construction work. However, my personal preference has always been as described below, step-by-step and I have always insisted as well as prevailed upon the project engineering personnel to design a precast rectangular monolithic RCC block as a common foundation base for all the structures, footings for all structures as well as for cable trenches rising from and resting on this monolithic foundation base at different locations, as required. My experience has been that such design has in most instances done away with pile foundations and has resulted in unimaginable fast completion of the entire substation and switchyard and often good saving of cost.

In case deep excavation (generally deeper than 4 or 5 meters) needs to be carried out in a large enough area, it is advisable to consider sheet piling in order to be safe from caving in of soil from unexcavated side towards the excavated area particularly because weather, the world over, appears to be getting more and more unpredictable and if by any freak chance an incident of unwarranted heavy rainfall occurs, whether or not some portion of the excavation will face caving in of soil of surrounding unexcavated area cannot be predicted. Experience calls for considering sheet piling in case project execution is on a site that is not virgin and whatever industry of plant is to be executed is in vicinity of an existing and operating plant, because seemingly innocent and unapparent vibrations are transmitted in the soil around foundations of most of the running machinery unless they are too insignificant – for example the existing industry pertains to tailoring of garments. However, if the soil strata at the site can, with full certainty, be considered to be exceptionally stable posing no risk of caving in, sheet piling cost should be saved.

CAUTION: Outdoor high voltage substations and switchyards though involve massive civil work basically fall in the territory of electrical engineering. From this point of view, I feel duty bound to caution the reader to be mindful of the fact that no matter how nicely has the civil works pertaining to these been carried out, the end result must comply with all requirements of electrical engineering. Fundamental requirement pertaining to outdoor high voltage substations and switchyards is efficient 'Electrical Earthing' or 'Electrical Grounding', which is supposed to be compliant of 'IEEE 80-2013 – IEEE Guide for Safety in AC Substation Grounding'. Therefore, any before those involved in design and engineering of civil works pertaining to outdoor high voltage substations and switchyards proceed with finalization of their designs, the matter must be discussed threadbare with the electrical engineering group with the purpose of ensuring the right 'Electrical Earthing' or 'Electrical Grounding' and achieving safe 'Step potential' and 'Touch potential' while establishing measurably good electrical contact with mother earth for allowing earth-fault currents to flow as desired in overall design and engineering of the system.

- In planning and deciding sequence of various civil works due care needs to be exercised in deciding areas for temporary safe keeping of dug out soil so that the same can be used for back filling after construction of civil foundations. Some dug up soil would become excess and would need to be disposed properly and as per the prevailing rules and regulations. Type and properties of such excess soil permitting it would be best to plan the activities in such a manner that to the extent possible the dug up excess soil is used for achieving grade levels.
- Method Statement / Risk Assessment for each and every activity should be prepared, in order to disclose what could cause harm, and respective due precautions, to be taken by people while executing the task, detailing therein the following:

(a) Methodology of execution – list if activities and procedures in sequence of their execution,

(b) List of machinery, tools and tackles proposed to be used for each activity or procedure,

(c) List of name of person or persons who are proficient in use and testing and inspection of the machinery, tools and tackles,

(d) Checklist of fool proof safety measures, points and precautions to ensure that the construction machinery proposed to be used is fit for the job from the point of ensuring the desired quality and accuracy and also ensuring freedom from hazards and accidents that may cause material damage or human injury/causality (obviously, schedules of maintenance and servicing of all construction machinery must be strictly adhered to and regular inspection and wherever required testing, of various vital appliances, apparatuses, tools, tackles, components and parts of such machinery must be properly inspected and tested at regular intervals and results of such inspection and tests recorded along with date, time, name of person and name of witness present).

While on the subject of Method Statement / Risk Assessments I would like to mention below a couple of incidents describing my own experiences in projects being handled by me.

*(i) A fatal accident, that was totally avoidable, occurred at a project site where a large power project was under execution on EPC basis by a reputed international contractor and I was project manager on behalf of its owner. Project site was at a location near sea (about 5 km) and soil at the site was black cotton type in nature needing very deep RCC friction piles, conforming to IS 2911-4, being driven in bulk.*

*As is the universal practice some test piles were driven and were supposed to be put under various tests as provided for in the said Indian Standard. I had not yet shifted my residence near the project site so as to be available there all through the working hours and was visiting the site once a week and also whenever necessary. When first of the piles was ready for test and I was on a site visit I happened to see the arrangement made by the subcontractor (a reputed civil contractor) of the EPC contractor and it occurred to me that whatever the arrangement was it was ill conceived as no really effective constraints were provided to prevent the heavy tetrahedron concrete blocks, of size about 1x1x1 meter, used for applying vertical dead load for testing load bearing capacity of the test pile.*

*I immediately contacted the project manager of EPC contractor and expressed my apprehension to him and suggested that (a) suitable highly effective constraints be provided for sake of safety and (b) suitable electronic arrangement for remote monitoring for measurement and recording of extent of sinking of the* test pile on application of the vertical dead load be provided so that in an unlikely event of failure of the safety constraints no body is hurt due to toppling of the heavy concrete blocks.

The EPC project manager kindly appreciated my suggestions and agreed to implement them all and advised me to rest assured that the pile testing work shall be carried out fully safely with all precautions. However, it was for the first time for the said project manager and he just fully and faithfully passed on my apprehensions to the project manager of the sub-contractor. This one act diluted the importance and possible effects of my observations and suggestions and what ultimately happened was that since the sub-contractor had to arrange the necessary electronic gadgets for observing and recording the sinking of test foundation under effect of dead loading, it was decided by the subcontractor that money shall not be wasted in providing the required constraints for ensuring that toppling of concrete block wouldn't occur. This one decision, for sure, introduced a weak link in the chain.

As a result the arrangement that finally was made was such that it required just a small trigger to cause a disaster and as luck would have it the person responsible for physically carrying out the test decided to carry it out very early in the morning – at 4.00 AM with the good intention that the test shall be over by the time the work would start at 8.00 AM. So far all went remained good. It can be anybody's guess what inspired the person who was carrying out the test to physically crawl under the platform, loaded with concrete blocks and kept over the test pile, to observe the needle of measuring instrument with his own eyes. God only knows what happened and which thing did this unlucky person touch that suddenly the platform toppled and the poor person was crushed to death under load of so many heavy concrete blocks each weighing about 7 tonnes.

Reading the above paragraph might have been difficult and I apologize for the discomfort but this incident is true and I wish that such incidents should never ever occur anywhere in the world. Needless to say that occurrence of a death at worksite creates legal complications that may bring all activity to a dead halt and lead to financial burden by way of compensations to be provided to the family/ dependents of the unfortunate deceased.

(ii)     Another fatal accident that I recall with remorse had occurred long back in the year 1999. The project pertained to addition of a coal fired thermal power generation unit to an existing power plant. The project site had hard soil with angle of repose of about 90o and digging of pits in soil in that area was customarily carried out maintaining angle of about 80o. The site was located in a tribal belt and Govt. regulations demanded maximum possible utilization of manual labour so as to provide employment to tribal men and women during construction work of industries coming up in the area.

In connection with the power plant extension work that I was handling a trench of width of 2.5 meters and depth of 1.7 meters was to be dug for a length of about 150 meters for construction of a RCC storm water drain. In accordance with applicable labour employment related rules the work was undertaken totally manually rather than using a shovel excavator. For ease of working stairs were cut in the soil at intervals of 20 meters for enabling female labourers to remove the dug-up soil by head loading in shallow aluminum pans.

It so happened that one location along length of the trench being dug up was a backfilled area that although must not have been adequately compacted at time of backfilling and didn't achieve quite the same extent of compaction and cohesion between soil particles as occurs naturally over centuries. As a result the slope of walls of the trench at his location was too steep and soil in this location suddenly caved in thereby submerging one soil digger totally. All the workmen and their supervisor who also was present rushed to the locations and started to remove the caved in soil by hands as fast as possible. They did not use shovels for avoiding hurting the man. However, it took about twenty minutes by the time soil above nose of the person could be removed. There was no sign of him breathing. Taking him out of the caved in soil took another about 20 minutes. Quite obviously the man died due to suffocation.

To my mind the only way for avoiding this incident could have *been use of shovel excavator rather than manual digging because it is not practically possible to go on testing soil characteristics for each and every inch of soil surface during such excavation work.*

*The lesson that I learnt from this unfortunate accident was that no matter what the idiotic labour laws and rules say, a project manager must respect his own hunch regarding use of machinery to carry out hazardous jobs for sake of human safety.*

- Before mentioning about rather known matters pertaining to safety which often get overlooked, it would be prudent to bring awareness about some problems which those who do not have exposure and experience of Indian scenario may not even expect to be occurring. The problem is that most of poor people, generally unskilled and semi-skilled workers (both male and female) are addicted to chewing tobacco. Problem with chewing tobacco is that anyone who indulges in it is forced to spit saliva from time to time because the such saliva cannot be gulped in the stomach generally by any human. Another problem is that a person who chews tobacco suffers from urge to continue doing so again and again because the kick that the addicted person gets dies out after a while and the person feels miserable if he doesn't take another does of chewing tobacco. It is well known that chewing tobacco repeatedly cause grave risk of the addicted person suffering from oral and/or throat cancer that ultimately results in his/her painful death. However, the core subject matter of this book does not pertain to solving such serious problems hence we leave the health and social aspects of chewing tobacco and proceed with its effect on industrial project execution.

The harms that the bad addictions of Indian workers bring to project execution are the following:

- Spitting produces grave risk of spreading of communicable diseases to others involved in execution work including persons of all levels and nationalities. If by any chance, some dangerous contagious disease is just spreading its tentacles spitting of saliva here and there in an uncontrolled manner may result into the disease taking epidemic dimensions. If this happens, it is but obvious that progress of project may get very badly affected.

- The urge to spit out saliva is so pressing and compulsive that a person who is chewing tobacco is compelled to expel the saliva per force due to reflex action of his/her body. Because of this it is not uncommon that people spit saliva even inside equipment that is under installation. I have seen persons spitting saliva inside steam turbine while diaphragms are under fitment in lower half of turbine casing. I can go on listing hundreds of such vulnerable locations of various machines, but one example should suffice.

It is therefore, but obvious that some highly effective method should be applied to prevent occurrence of spitting of saliva within project site by anyone. The foolproof method is putting into regular practice of bodily search of each and every worker, irrespective of his/her rank, by Security Personnel so as to ensure that none of them is able to carry with himself/herself any of the following:

(a) Pouch of chewing tobacco or other tobacco products,

(b) Pack of cigarettes (in India apart from conventional paper rolled cigarettes, an indigenous version of it known as 'Bidi' is very popular – security personnel know about it),

(c) Matches or lighters (as both these would be of no use unless someone indulges in smoking and can be source of breaking out of a fire incident).

- Highly functional and unfailing mechanism for double check of all safety measures must be provided and all such checks must be recorded in real time with name and signatures of concerned personnel along with date and time when the double checks were carried out. Any special observations made by the checking personnel should also be documented.

- Appropriate Information System should be put in place, making it possible to provide highly efficient use of cellular phones for ensuring real-time flow of all information by way of development of APP that will allow various people to log on to a common database resident on a server for making their respective entries on various tables/ pages so that there is created a tool that gives audio signal and/ or vibratory signal, of occurrences of new entries regarding various activities of project team members, to the project manager on his/her cellular phone thus becoming his/her real-time ears and eyes about dynamics of project activities.

- Another most important preparatory work is deep study of the execution schedule and listing down of requirements of all the Assets needed such as Cranes, Excavators, Backhoe Loaders, Bulldozers, Skid-steer Loaders, Graders, Tracked Loaders, Trenchers, Wheel Tractor-scrapers, Dump trucks, Trailers and Concrete Pumps etc. of various types and capacity and periodicity of their requirements. This information may be used in proper and economical sourcing of such construction machinery, planning their parking spaces, placement and required maintenance spaces and management of consumables for them. This preparatory work can be of real benefit in locations such as notified Industrial Areas, Industrial Parks and Special Economic Zones where several industries of different types may come under execution simultaneously thus resulting big requirement of construction machinery and difficulty in their timely and economical availability for hire from the parties who are in the business of supplying such machinery for project construction. Also, please note that it is not advisable to totally rely on a single source of supply of these assets. From this point of view, it is advisable to locate at least two sources and to tie up with both of them to supply these assets in parts. Doing so is likely to be helpful in ensuring trouble free availability of the assets if some unforeseen problem arises with one of the two suppliers. For very large projects with execution time of more than three years it may be useful to tie up with more than two sources of supply of these assets.

- It is equally important to ensure that before being put to use at project site each construction machinery has been duly tested for endurance, stability and safety. Subsequently, it should be assured that regular testing and inspections are carried out by competent persons and records of testing and inspection maintained properly. Availability and accessibility of such records too must be ascertained.

- Similarly, source of availability of basic civil construction materials such stone aggregate and sand of the desired type (of proper type suitable for making concrete that is free of alkali-aggregate reaction and is of the desired strength), may be located and appropriate action taken to assure adequate and timely and adequate availability of these materials at the site, alternatively reputed and reliable source of premixed concrete of the desired type (it is highly advisable not to just believe the supplier in good faith but to arrange testing of the materials at two proper independent laboratories, one of which should preferably belong to a reputed engineering institute or college, so as to cross check the results – it would always be very useful to always suspect all testing laboratories except those belonging to the premier engineering institutes known as Indian Institute of Technology - IIT).

For sourcing the basic civil construction materials or premixed concrete it is important to tie up with at least two alternative sources or suppliers. If premixed concrete is the choice, matters may be relatively less complicated if all due quality control procedures are strictly adhered to. But in case stone aggregate and sand are being sourced from quarries, it is also extremely important to ensure that the material that reaches the construction site is the one that was mined and loaded on the transport truck at the quarry – please do not assume that the truck loaded with good quality material will directly reach your site without delivering your material to someone else and getting reloaded with low grade material that may ruin your civil construction work.

## *Starting physical work at construction site*

It would go a long way in ensuring smooth execution of the project with all due cooperation of the workforce if just a few local rituals and customs are adhered.

There is a famous saying, "When in Rome do as the Romans do". It means to say that having to work in a foreign land, follow the customs of those who live there. It can also mean that when you are in an unfamiliar situation, you should follow the lead of those who know the ropes.

In India people believe in powers of the supernatural or God (and there is just not one God instead, there are innumerous Deities of Hindus and which Deity to worship depends upon the State where the project is situated) and it is customary here to start every new work with worship of a God or Deity.

When the first physical construction work, even as menial as start of earth digging for some work, is going to be undertaken, irrespective of the number of labourers that would be engaged in this work, it is highly advisable to arrange performing of a worship (generally called 'Puja' in most parts of India). In your work team, there should be a middle level supervisor who should preferably belong to the area where the project is located or at the least the person must belong to the corresponding state. This is needed with the view that such a person will be able to arrange a priest for performing the Puja and would be able to arrange the appropriate. The preparation will require arranging local good quality Indian Sweets, Coconuts (for being broken as offering) and hygienic drinking water. The Puja may last at the most two hours and shall end with distribution of sweets and pieces of coconut to the labourers and all others – including non-Indian members of the team and visiting officers. This small gesture will enable bonding of the labourers to the project and your organization. Such Puja should be performed each time some new work is started. Of course, the expense to be incurred during each Puja will go on increasing as the number of labourers engaged at the work site go increasing from tens to thousands and advance action may be required to arrange the sweets for distribution at end of each Puja performance.

Each such ritualistic performance of Puja may disrupt the work for about four hours. Thus, if say 100 workers are engaged, each Puja may mean loss of man-hours to the tune of 400, which certainly is not desirable as strength of workers engaged increases from hundred to thousand. Easy trick for avoiding such gross loss of man-hours is that as the area of work and volume of work increases with progress of project, the project area is divided into several seemingly autonomous departments, each such department having strength of at the most a hundred or so workers and technicians. Doing this would minimize the loss of working man-hours in performance of the ritualistic Puja, as each of the ceremony would then pertain to just one department of work area rather than to the project as a whole. It is but obvious that expense towards distribution of sweets and coconuts for each Puja would reduce substantially.

Simultaneously, arrangements should be made for free of cost and adequate availability of hygienic drinking water (generally cool, but not about 20 C temperature) and a subsidized place with simple low cost seating and eating arrangement where the labourers can go for buying eatables (vegetarian snacks that are customarily consumed by common people in the area where project is located) and tea (called chai) and where sumptuous but hygienic and tasty vegetarian lunch (because it is not possible to ensure that non-vegetarian food is healthy) is available on a no-profit-no-loss basis or at a subsidized cost, as may be desirable.

Such a place for serving snacks and food is customarily called

'Canteen' in India. It would serve as 'icing on cake' if the project manager shows his face in the Canteen, during lunch hours, once in a while but someone senior enough from the managerial team of project does make it a point to show his face in the Canteen, at lunch hours for eating what is being served to the workers. Eating together forms strong human bondage between people in India therefore this small gesture would create a feeling of indebtedness to project management organisation among work force.

Having mentioned about psychological, social and welfare aspect of starting physical work at site I would like to stress upon importance and necessity of establishing a facility that may be named "Safety Training Hall" and should have size to enable comfortable training of about twenty persons.

The Safety Training Hall should have a toilet nearby and should at the least be equipped with the following:

- A White Board (preferably magnetic) along with a set of White Board Pens of various colour, a Dry Wipe Cleaner (preferably magnetic).
- A colour LED TV suitable size for projecting training material.
- A Desktop PC with matching Computer Power Inverter and big enough battery and Surge Protective Device.
- Safety Training Software such as small videos, Powerpoint Presentations – all preferably in local language or alternatively translated by a competent trainer scene-by-scene or slide-by-slide as the case may be.
- Software for recording Training Progress of each candidate.
- Necessary furniture including good illumination and ventilation. If possible providing Air Conditioner

should be avoided. At the most arrangement may be made for evaporative cooling during summer and water from its storage tank be drained off daily at end of the day so as to prevent breeding of malaria causing mosquitoes. Providing Air Conditioned comfort for personnel of any level at most of the Indian project site, in my personal experience, is counter productive because (i) contrast between temperature prevailing in Air Conditioned space is far too low as compared with that prevailing at the work site and exposure to such sudden change of temperature either from cold to hot or vice-versa is harmful for human health and (ii) the experience of comfort during an hour or so of sitting in Air Conditioned Comfort will have an unpleasant psychological effect on the workers who will repeatedly get the experience during training sessions and their minds will be forced to compare themselves with their supervisors and senior officers looking after the project and will resent their luck each time the quality or extent of work done by them is criticized and they are given sermons about properly doing their work because they will say, in their hearts – it's easier to preach than to do. History has it that all successful generals who have won difficult wars have faced hardships of the battlefield side by side with their soldiers.

A project site throws a battlefield like challenge to its project manager since in a battle there normally are no schedules to be met either in terms of time or in terms of cost of winning the battle while in total contrast from success in a battlefield the crux of success in execution of a project is measured in terms of meeting the time schedule while not only not exceeding the estimated costs but also saving on it. Therefore, project manager who succeeds is the one who is wholeheartedly supported as well as regarded by each and every member of his team. No doubt project manager is tasked with setting examples of determination, tolerance, clarity of vision (never having to cut a sorry figure due to a work having to be redone because of either an error in drawing or due to having been scheduled wrongly or some other reason – whatsoever).

- Safety Training sessions should be no longer than about 50 minutes (most human brains, in general, fail to register whatever is told verbally for more than 45 to 55 minutes) and each session should end with provision of tea to trainees as also by the trainer – who should indulge in some friendly chitchat with the trainees. Such Safety Training sessions should be repeated, each time followed by an oral examination of each trainee for assessment of extent of learning achieved. After few sessions, there can be held a paper & pen or IT based test comprising objective questions to be answered by tick marks. While on the subject please note that best time for imparting training is between 10 AM and 3 PM. Holding training sessions after 3 PM are likely to be wasteful and must be avoided.

- Marks obtained in such paper & pen examination should preferably be communicated and the first three rank holders should be given a small token gift (which can even be a box of chocolates) as a mark of appreciation.

- As the project gains momentum, the number of persons to be imparted with Safety Training will obviously go on increasing since each worker has not only got to be trained but also retrained until it becomes clear that he/she has really grasped what has been imparted to him/her. This makes the task of trainer quite tedious as he/she has to go on holding training sessions one after the other in quick succession. Therefore, a point of time may be reached when more than one trainer may have to be engaged.

- Safety Training course coverage should include the following.

    - Common Hazards at work sites of industrial projects and how to safeguard against them all.

- Project Specific Safety Hazards, general and work area specific and how to protect oneself from them.

- The right dress for men and women working at a industrial project site, with due consideration to what men and women who work as labourers like to wear while at work and merits and demerits of it.

- Basic Protective Gear and their importance – Safety Shoes and how to wear them, Helmet and how to use it, Safety Gloves and where and how to use it.

- Special Protective Gear specific to work areas and nature of work to be done – Safety Goggles, Electrically Safe Safety Shoes, Electrically Safe hand gloves, Protective Apron and its use and similar other safety gear, Safety Gear to be used for working at heights.

- Health and Safety Signage and their meanings.

- Dos and Don'ts in case of occurrence of an accident. It should cover, in particular, what to do if someone is himself/herself meets with an accident (needless to mention that the project manager must provide a fool proof mechanism of promptly bringing to notice occurrence of any accident in any area).

Given above is not an exhaustive list of what all is required because it would vary from project to project. But I do hope that the above serves to put the matter of Safety Training in right perspective.

- Safety Administration: Indians in general tend to take matters for granted and are deep rooted in short cuts – often to save time or to avoid minor discomforts and therefore habitually indulge in avoiding use of safety gear and abiding with safety rules and regulations. People not only traveling in motor cars but even those driving motor cars can be seen not using safety belt. People can be seen not using Driving Helmet while driving two wheelers.

Even highly educated persons just don't have any idea about the following, with reference to Safety Gear – in relation with driving motorcycle.

- Visors or goggles to protect eyes from wind, rain, insects and road dirt – hardly any one wears safety goggles or visors while driving a motor cycle, perhaps due to lack of knowledge coupled with financial limitations;

- Safety Helmets – for wearing safety helmets there are Government Regulations but climate in most of India is such that wearing safety helmet results in heavy perspiration above neck and created a suffocating feeling (may be motor cycle is not a vehicle that should ordinarily be used by any body in India but economic conditions have made two wheelers the most common vehicle of the masses);

- Safety Clothing – generally these are specifically meant for protection from cold and wet weather both of which do not exist in most of India but even areas where weather is cold all around the year (Himalayan region) people don't have any idea; although the situation is also linked with poverty prevailing in the country;

- Gloves – people just don't know that gloves should be worn while driving motorcycle;

- Boots – people rather like to wear sandals;

- Visibility aids.

My purpose of giving the example of tendency of avoiding use of safety appliances while driving motorcycle is just to give a glimpse of what the scenario may be at the project site, if matters related with safety are left to the workforce without making really fool-proof arrangements to ensure that no one engaged at any work at project site is knowingly or unknowingly able to indulge in practices that are unsafe. Obviously, this calls for not only imparting training but to also put in place a mechanism of detection of indulgence in unsafe practices. Someone may think that this amounts to putting a policeman behind each worker but there are other ways too which we will discuss as we proceed further.

However, I feel that it would be appropriate to give an example of what really fool-proof arrangement is made in case of high voltage transmission line towers all over the world including the countries to which maximum number of Nobel Prizes pertaining to physics have gone. Each transmission line tower is provided with an anti-climbing device that makes it impossible for anyone to climb a transmission line tower beyond a certain height.

- Monitoring to ensure that use of Safety Appliances is not compromised anywhere at the project site. I don't really think that I need to include a write up on this subject because in modern times monitoring anything is perhaps the easiest thing to do. However, with a view of not leaving this subject untouched, I would like to make a mention of extensive use of CCTV Cameras of good quality duly wired to enable close-up viewing in real time and hooked to good enough a number of high resolution monitors in a control room wherein there are several observers keenly observing what goes on in the project site. As project progresses the plurality of CCTV Cameras, Monitors and Observers will need to be increased and location of CCTV Cameras would need to be changed from time to time. This would be dynamic and if the project progress is really good such changes would be needed repeatedly. Needless to say, that CCTV Cameras should also be appropriately made use of not only for safety but also for security and for proper watch and vigilance of the site during night.

- Giving monetary incentive to workers who display exceptional adherence to safety procedures. I have experienced that even though workers may not be directly employed by the organization of project manager, it pays to provide small incentive to the workers to strictly adhere to all safety requirements and to always remain mindful of possibility of accidents during working or even during presence on a work site. The incentive can be in terms of an award of a small sum, for example, Indian Rupees (INR) 100/- (one hundred only) to all workers who do not meet any accident, however minor (for example needing nothing more than applying some antiseptic on a bruise) during a calendar month and special award of additional INR 1,000/- to all workers who manage not to meet even most minor or unreportable accident for a whole calendar month. Workers really care to be able to get such awards and try to make project site free of accidents. Such monetary or tangible small rewards (e.g. certificates, phone cards) are to be distributed to the winners during weekly site safety meetings welcoming others to invest their time and effort to improve their safety and encouraging workers to do the same.

- Security and Surveillance of the Site.

Before any work can really be started at site it is essential that the site is made secure enough and adequately impervious. This has got to be done by the party whose responsibility it is to do so depending upon corresponding scope as provided for in the contract. However, the following is meant for giving a good enough idea of what is desirable.

In the following is an indicative list of all that is generally considered as requirements towards making a site secure and impervious, although the following is the bare minimum requirement rather than comprehensive and much more may need to be done depending upon the state of affairs in the state where the site is situated and law and order situation at the site proper.

- Strong, impervious and high enough boundary

wall, with anti-climbing protection (electrical fencing is illegal in India). There should be no holes in the boundary wall for any reason or purpose.

- On inside of boundary wall, all along it, there should be a motorable road of a minimum 5.5 meters width and with hard enough surface so as to allow a patrol vehicle to run irrespective of weather conditions. There should be provision of good unfailing illumination all along the inside of boundary wall. Needless to say that good enough storm water drains should be provided so that patrolling may not be affected by water logging during heavy downpour in monsoon season.

- Effective Security Surveillance system using modern technology should be provided all along the boundary wall so as to enable prompt detection of intrusion.

Now in India high technology driven security and surveillance systems and equipment is available. Indicative list of solutions that are available is the following:

1. Perimeter security (using buried or above the ground sensors and radars, high resolutions & thermal cameras),

2. Integrated gate checkpoint for ensuring controlled entry and exit to and from site. The system may include biometric identification at entry to and exit from site, gate/door automation, under vehicle scanner, metal detectors, baggage scanners, high-resolution surveillance cameras etc. As time passes from the date of authoring of this book much more advanced and developed equipment is likely to become available and there can't be any reason why the latest should not be implemented.

3. In-building security including high-resolution surveillance cameras, fire detection, access control solution, public announcement system, intrusion monitoring system etc.

Usually the parties providing these equipment and services don't provide manpower for observation of monitors hooked up to CCTV cameras but they do provide resident engineers for Operations & Maintenance if installation is large enough.

It shall be helpful if Motion-detector activated Flood Lighting and Siren are also added for further improving security.

Wireless communication system having a sufficient range and comprising good enough plurality of handsets should be provided for effective communication between security personnel (just relying on cellular network for security related communication is not advisable as there can be umpteen number of reasons for cellular network not being available just at the moment an intrusion occurs at site and security personnel urgently need to communicate.

Through a local officer rapport may be established with the local police department and regular friendly interactions may be maintained. However, it would not be wise to rest assured that just some good rapport with the local police shall make site secure enough because you never know – a blood relation or even a real brother of some policeman may be a gangster. Therefore, rapport and good relationship with local police personnel may not be construed to be a guarantee for security. For this purpose, Indian origin security officer of the project has to be a shrewd, intelligent person with sharp-wit and good far-sight. Often retired military personnel of the rank of Brigadier, Colonel and Lieutenant Colonel are provided by Security Manpower Agencies. It may be advisable to set-up the Security Department by engaging such personnel who may form a harmonious team. However, post appointment their performance needs to be watched for a couple of months and there should be no hesitation in pressing the Security Manpower Agency to replace any person who is found not be up to the mark so far as performance is concerned.

If the project is big enough and is likely to take time of more than a year then it may also be advisable to develop a setup of smart and healthy enough retired Govt. Officers of second rank for the purpose of establishing rapport with local administration (Commissioner, Police Superintendent and Civil Surgeon) whose help may become necessary, should some criminal incident occur at site. As a matter of habit a project manager should never take things for granted – there can be many slips between cup and lips, so far as Security is concerned since matters are very much human driven. Nonetheless, there is nothing to be scared of as generally project sites are never infested with hardcore criminals and problems arise generally from small criminals. Recommendations have been made above for implementation of high technology gadgets for the purpose of surveillance and security as these are certainly likely to payback for the expenses should, by any unlikely chance, there be some attempt of a sabotage.

Irrespective of whether or not security and surveillance falls in scope of the party implementing a project, it is highly recommended that a good and strong enough barbed-wire fencing with proper lockable gates at entry and exit points of project area, falling under scope of implementation of project, be constructed and good enough security and surveillance equipment be installed. However, in such cases it may not be necessary to establish rapport with local administration.

## Management of Civil Work

Once site has been made ready for starting execution of project the first broad activity that needs to be undertaken is civil works, key to success in timely execution of which is time of start being appropriate. Much more effort, money as well as time is likely to get wasted if work is started when time is not opportune. Except in a couple of areas that are arid or deserts rains are for long hours and often for days together. As a result, it may sometimes become to redo a civil work if nature of the work is such that heavy downpour for long hours is considered detrimental to its quality.

What is most desirable is that site is taken over, by the party who is responsible for execution of project, at a time when monsoon has just ended. This way it can be assured that by properly sequencing and scheduling various activities good enough progress can be achieved when next monsoon approaches. With well thought out sequencing and scheduling of work it can be so managed that on arrival of monsoon season almost all the foundation work including backfilling of soil around foundations is already finished, as well as due accessibilities are granted, and the work that needs to be carried out during monsoon season is all above ground level and out of soil.

In spite of such good sequencing and scheduling of work the work may have to be allowed to come to a halt in certain states in North East (for example Assam) where rainfall is almost always incessant and recurrent and even transport of goods and machinery becomes near impossible not only because of rainfall and flood but also due to landslides. In such situations, it may be best to let it come to a comfortable halt and to utilize the period of monsoon for working out plans for the period when monsoon comes to an end.

With the knowledge as above let us assume that all has been

well planned and chances of disruption owing to monsoon is ruled out as work is to be carried out just after end of monsoon season. Certain special precautionary actions if taken are likely to prevent losses of time and money due to necessity of rework owing to quality problems and errors.

**For large volume concreting activities:**

Let us recall Murphy's Law, "If anything can go wrong - it will" or "Whatever can go wrong - will go wrong". This is the first and foremost of all precautionary consideration that should almost become a habit of a project manager (but not to the extent of a psychological obsession).

In the following I have narrated my experience of two instances where things had actually gone wrong:

Reinforced cement concrete (RCC) (also called reinforced concrete or RC) is a composite material in which concrete's relatively low tensile strength and low ductility are counteracted by the inclusion of reinforcement having higher tensile strength or ductility.

The reinforcement is usually, though not necessarily, steel reinforcing bars (rebar) and is usually embedded passively in the concrete before the concrete sets, meaning thereby that rebar are duly shaped and placed in the desired positions and locations prior to pouring of concrete. Reinforcing schemes are generally designed to resist tensile stresses in the structure in particular regions of the concrete that might cause unacceptable cracking and/or structural failure. Modern reinforced concrete can contain varied reinforcing materials made of steel, polymers or alternate composite material in conjunction with rebar or not.

Reinforced concrete may also be permanently stressed (concrete in compression, reinforcement in tension), so as to improve the behaviour of the final structure under working loads.

123

Such reinforced concrete work is known as Pre-stressed concrete (PSC). We will confine our discussion here to RCC because PSC is generally not used for construction of industrial projects, PSC being dominant in projects pertaining to large bridges.

Obviously, therefore, the matter of first and foremost importance is ensuring the right source of reinforcement steel bars (or rebar). In India, the most common rebar material is Thermo-Mechanically Treated (TMT) Steel Bars, properties of the steel of which the TMT rebar are to be used is decided by engineering. Project manager is tasked with ensuring sourcing the TMT rebar material of the right quality and ensuring that the quality is consistently maintained. This is quite a herculean task for TMT rebar material is actually delivered to project site by authorised suppliers linked with various manufacturers.

Along with very good manufacturers of steel there also are those that produce TMT rebar from scrap material (generally linked with ship breaking industry). A list of reputed steel manufacturers is given under "Management of Steel Structural Work".

The problem with TMT rebar produced from scrap material is that their alloy consistency is not maintained. Such rebar produced from scrap material are often embossed / stamped with symbols identical to those of high quality manufacturers and the so-called authorized distributors try to push such spurious TMT rebar to unsuspecting customers. They even provide Mill Rolling Certificate of the manufacturer of genuine high-quality steel.

The problem becomes really acute in cases where project execution site has saline subsoil found all along India's coastal areas calling for use of corrosion resistant TMT rebar.

India's most famous Tata Iron & Steel Company (TISCO)

manufactures corrosion resistant rebar under brand name 'TISCON CRS 500D". It is manufactured under controlled processes through ultramodern machinery using latest technology. Others manufacture corrosion resistant steel rebar conforming to Bureau of Indian Standard (BIS); Fe 500 and Fe 500-D.

The following table shows a comparison between BIS and TISON product.

| Comparison of specification of TISCON CRS 500D with Bureau of Indian Standard Specifications for Chemical properties of corrosion resistant steel rebar | | | |
|---|---|---|---|
| **Chemical Properties** | Fe 500 | Fe 500-D | Tata TISCON 500 D |
| % Carbon | 0.300 | 0.250 | 0.250 |
| % Carbon Equivalent | 0.420 | 0.420 | 0.400 |
| % Sulphur | 0.055 | 0.040 | 0.035 |
| % Phosphorus | 0.055 | 0.040 | 0.035 |
| % Sulphur & Phosphorus | 0.105 | 0.075 | 0.070 |
| % Nitrogen (PPM) | 120 | 120 | 120 |

It therefore becomes a matter of prime importance to ensure that in the name of TMT rebar the material that is delivered to project site is genuine and consistent in quality, irrespective of whether the choice is rebar conforming to BIS or TISCON 500-D.

Best that can be done to ensure this is to discuss with a good steel manufacturer the feasibility of direct delivery of the TMT rebar to project site without involvement of the authorized distributor (doing so may become possible if the commission that the authorized distributor gets is agreed upon to be paid through corresponding paper work although the carrier truck loaded at steel manufacturer's work travels directly to project site to deliver the material). Even if this arrangement is agreed upon project manager cannot afford to assume that all will go as planned and it would be best if a person from project management set-up accompanies the transport vehicle all through from manufacturer's works to project site lest the transporter may offload good quality rebar at some place and get the carrier truck reloaded with spurious material.

Above is just an example and a suggestion. Having understood the situation and possibilities of undesirable rebar being delivered in place of the good quality ones, project manager may device his/her own ways and means of ensuring that only the rebar of quality as desired are delivered to the site. Managing it this way may be a necessity irrespective of whether the choice of manufacturer is Tata Steel or anyone else.

In the following is described another incident related with Concreting at a big project site. A lot can be learnt from it.

Concrete Batching Plant (CBP) of the civil works sub-contractor failed midway during bulk concreting as its electrical motor burnt, an identical spare of which was not available at site, also was not available a second CBP within the site. The civil works sub-contractor's site manager immediately rushed to another project site in the same area for possibility of getting concrete from the CBP available at that site, in spite of my objection to use of concrete from that CBP since the nature of concrete from that site was unknown and concrete was neither tested for use at my site from the point of view of freedom from Alkali-aggregate Reaction nor for Concrete Mix Design for the strength needed. Clock was ticking as both – the EPC project manager as well as the site manager of the civil works sub-contractor were perspiring, so was I. A maximum of about 30 minutes of time is allowed for a cold joint in concrete and it was becoming but obvious that the half-way done concreting was going waste and project was inching towards delay and but obvious financial losses to some party – either the EPC contractor or the civil works sub-contractor because redoing the work necessitated not only breaking the concrete that was laid till the incident but also the reinforcement steel bars. Needless to mention that the losses did occur.

The entire mishap could almost certainly have been prevented had someone (either EPC project manager or site manager of civil works sub-contractor would have devoted himself to think what could go wrong and taken action to mitigate the effect of it). I have learnt the hard way that for each concrete laying activity at least one hundred percent alternative must be available duly tested and rehearsed.

If the concreting volume available is low two concrete mixers of required capacity (for example 1000 liters) should be placed at strategic locations, both close enough and both should be made fully operational and tested prior to starting of the concreting. It may at times be advisable to also have required quantity of cement bags, stone aggregate, sand and water ready for both the concrete mixers. If the concrete mixers are electrically driven it is also advisable to have one of the two concrete mixers fully hooked up to a Diesel-generator of adequate capacity duly ready with Diesel fuel already filled in fuel tank and tested for operation after having being hooked-up. If, however, the concrete mixers are both Diesel engine driven then fuel tank of both the concrete mixers should be filled with adequate quantity of Diesel fuel and engines of both should be started and run for some time to prove them.

If, however, the volume concreting work is large, it is but obvious that the required concrete would be obtained either from local CBP installed at the site or from pre-mix concrete suppliers. In such a situation, it is advisable that project manager better carry out a brain storming exercise among his/her project team members and come to a proper conclusion about what can be the best foolproof arrangement for ensuring that an incident of disruption of supply of proper concrete midway during concrete casting does not occur.

Some of the measures that I have experienced are:

In case project site is large enough and there are more than one civil contractors/ sub-contractors each having its own CBP installed at site, it is best to make one of the two CBPs as a hot standby for the other. When I say 'hot standby' I mean that the standby CBP is really fully ready with all the needed manpower, material (stone aggregate exactly of the required type, cement and sand all in adequate quantities so as to start delivering the required concrete positively within fifteen or twenty minutes of being notified. For it being possible, both the CBPs should be duly properly connected to electrical power supply and motors of the standby CBP should have specially been tested, especially for the concerned concreting event, to prove that up to the motors all is working well. Thereafter, rest of the arrangement of the standby CBP should also be tested for being operational. This type of checking is absolutely necessary and the fact that the CBP to act as standby for the even had worked the previous day must never be considered to be a guarantee for its being fully healthy and fit for coming to rescue when needed in emergency. Needless to say, that at least one spare concrete transport mixer must also be kept at site as standby because concrete transport mixers are known to breakdown and it always pays to presume that any machinery or device or service that is known to have ever failed anywhere will certainly fail during the concreting event under consideration.

Equally important is to arrange one spare concrete boom placer, which should be kept in a state of readiness for being put to use with no loss of time if the one of those already in use does suffer a failure of some kind. Here the meaning of, "readiness for being put to use" implies that the spare concrete boom placer should be duly made ready with its outriggers fully deployed and it having been tested for reach being as desired. It is but obvious that proper geometrical study should be done to determine the right location for placement of the spare concrete boom placer.

My experience is also that with appropriate negotiations making such an arrangement that both CBPs are simultaneously in operation and supply concrete to the location of concreting activity. When I have managed this I have always planned the work in such a manner that ultimately when the concreting work reaches completion concrete boom placers of both concrete feeding arrangements meet so that even at end of the work there occurs no cold joint in the concrete work.

It is equally important that it is presumed that electrical power supply will definitely fail mid-way during the concrete laying activity and to mitigate it not only one but two Diesel-generating Sets, each of adequate rating so as to suffice singularly, should be mobilized, both duly ready with adequate quantity of Diesel fuel filled in respective fuel tanks. Both electrically hooked up to each of the two CBPs and the electrical hooking up duly tested and proven to the extent that in the event of an electrical power failure from the mains, all that is required is putting OFF the switch or circuit breaker that connects the CBP to mains power supply and then putting ON the switch or circuit breaker that brings power supply from the DG set meant to supply power to the CBP.

In case concrete is sourced from a Ready-Mix Concrete (RMC) supplier, concerned manager of the concreting work in the project, be it site manager of a civil sub-contractor or main project manager, should visit the RMC supplier's plant for assessing not only the adequacy quality control measures but also to make sure that the RMC manufacturing plant is such that a disruption of supply of RMC would not occur midway during the concreting activity at project site. It is highly advisable that free and frank discussions are made with the RMC supplier's plant manager and tough questions should be asked should it be found that there does exist some weak link in the chain. Crux of the matter is that it must be made sure that under no conceivable circumstances a disruption of supply of RCM will occur.

Having done the due diligence as above and made sure that the RMC supplier's plant is really capable of non-stop delivery for the concreting for the activity. This should include availability of adequate plurality of Concrete Transport Mixer or Truck Mixer (TM). Judgment of adequacy of availability of TMs in the required plurality should not be made subjectively but should be arrived at via a study and corresponding computations considering distance between RMC supplier's plant and project site, average speed of TM depending upon road conditions, conceivable traffic blockages such as railway track level crossings, traffic signals on the road and time taken at RMC supplier's plant to load each TM.

Job of manager concerned doesn't end with the above. It is also necessary to find out if there exists an alternative route with road good enough for the TMs to properly travel to and fro the site should it become necessary to take that route due to some totally unimagined circumstances. It is not possible to list out such circumstances in this book simply because whatever can be listed out has already been imagined.

If there are more than one RMC supplier in the vicinity it is but obvious that while choosing the supplier negotiations would have occurred with more than one RMC supplier. If this is the case, it is highly advisable to discuss with one of the other RMC suppliers the possibility of delivery of RMC to the project in case of a total disruption of delivery of RMC from the chosen RMC supplier.

## Brick Masonry Work

Other than concreting there can be brick masonry work, which is relatively simple. However, the bricks used in India are solid and made either of burnt clay or fly ash obtained from coal fired power plants. Standard dimensions of brick in India are 9 inches x 4 inches x 3 inches. While Indian Standard IS 1077

(1992) provides standard specification for burnt clay bricks; IS 5454 pertains to 'Methods of Sampling of Clay' and IS 2212 (1991) pertains to 'Code of Practice for Brickworks'.

Biggest problem with burnt clay bricks that are available in India is Crypto-florescence. This may result in bursting off of outer skin of bricks and is a result of setting in of creep of clay. Properties of burnt clay bricks and their behaviour depend upon (i) clay, shale or earth of which bricks are made, (ii) kiln atmosphere conditions, (iii) properties of water used for manufacture of bricks, (iv) mortar used to lay bricks and (v) materials with which bricks may come into contact with in wet weather. British Standard BS 3921: 1974 mentions about brick properties such as Soluble salts, Clay type, Irreversible moisture expansion, Water absorption, Initial rate of suction and Compressive strength.

Obviously, variables are too many – hinting to choosing some alternative even though a bit expensive, if it can provide freedom from complexity of undesirable properties burnt clay bricks, unless someone finds a highly reliable source of burnt clay bricks that can be considered to be capable of lasting long enough – preferably life time of project concerned.

An alternative available in India is Fly ash brick (FAB), containing class C or class F fly ash and water. Compressed at 28 MPa (272 atm) and cured for 24 hours in a 66 °C steam bath, then toughened with an air entrainment agent, the bricks last for more than 100 freeze-thaw cycles. Owing to the high concentration of calcium oxide in class C fly ash, the brick is described as "self-cementing". The manufacturing method saves energy, reduces mercury pollution, and costs 20% less than traditional clay brick manufacturing.

A less expensive cement known as FaL-G (Flyash-Lime-Gypsum) cement was introduced and patented by one Dr. Mrs. Bhanumathidas and Mr. Kalidas in 1991. This revolutionized fly ash brick manufacture in India. FaL-G technology (www.fal-g.com) has simplified the process by adding gypsum to fly ash + lime/cement, converting the calcium aluminates into calcium alumino-sulphates resulting in to achievement of high early strength. Thus FaL-G brick does not need any pressure and gets cured at ambient temperature of 20-40 °C. By avoiding both press and heating chamber, FaL-G process has brought down the multi-million Rupees brick manufacturing plant cost to a few hundred thousand Rupees, making it within the reach of micro units. This facilitated proliferation of over 18,000 FaL-G Brick manufacturing units in the country as of 2016 and many more must have come up thereafter.

## *Alternatives to conventional natural civil engineering materials in India*

Saving in terms of cost and time are the crux of successful execution of industrial projects. From this point of view, in case contract pertaining to project doesn't specifically prohibit use of alternative civil engineering materials, it may be worthwhile looking for alternatives in the area where project is situated. However, alternatives to conventional civil engineering materials should be ventured into only for civil works of relatively lesser strategic importance such as paving and concreting for road construction while such alternative materials may not be considered to be used for main and important civil construction works such as foundations and structures. Moreover, possibility of using alternative civil engineering materials may, in general, not be even looked into unless project manager himself/herself is a well-qualified and highly experienced civil engineer having prior knowledge of case histories of usage of alternative civil engineering materials in various project preferably in India or elsewhere in the world or alternatively project manager is in direct consultation with an acknowledged expert on the subject (caution is advised to ensure that the person supposed to be the expert in alternatives to civil engineering materials in India, is neither a sales agent nor is expected to be a beneficiary in case he/she is able to convince the person under advice to make use of such alternatives. However, if project manager is fully knowledgeable about civil engineering materials and methods of testing their characteristics, quality and usability there may be merit in looking for alternatives if the right ones are available in the particular area at attractive price.

Today Indian Standards are copiously used for ensuring quality of construction of buildings and other structures, which are nowadays largely dependent on concrete constructions. Bureau of Indian Standards, the National Standards Body of the country, considering the scarcity of sand and coarse aggregates from natural sources, has evolved number of alternatives which are ultimately aimed at conservation of natural resources apart from promoting use of various waste materials without compromising in quality. These measures include permitting in the Concrete Code (IS 456) as also in the National Building Code of India, the use of slag - a waste from steel industry, fly ash - a waste from thermal power plants, crushed over-burnt bricks and tiles - waste from clay brick and tile industry, in plain cement concrete as an alternative to sand/natural aggregate, subject to fulfilling the requirements of the Code. This Code, further, encourages use of fly ash and ground granulated blast furnace slag as part replacement of ordinary Portland cement in plain as well as reinforced cement concrete. This part replacement could be of the order of 35% and 70% for fly ash and slag respectively thereby affording a large scale saving of natural limestone reserves, which would otherwise have depleted in case of the use of ordinary Portland cement without such replacement. Not only this, the Code highlights how durability of concrete can be improved with the use of these supplementary cementitious materials. The Indian Standard on concrete mix design (IS 10262) has been upgraded to include guidance and examples of designing concrete mixes using fly ash and slag. Provisions for compliance for requisite quality of concrete made using fly ash and slag have been duly covered for the manufacturers of ready-mixed concrete in the Indian Standard Code of practice for RMC (IS 4926).

BIS has also formulated an Indian Standard Specification for artificial lightweight aggregates covering manufactured aggregates, such as foamed blast furnace slag, bloated clay aggregate, sintered fly ash aggregates and cinder aggregate (IS 9142). A series of Indian Standards has also been formulated on various precast concrete products such as solid and hollow concrete blocks, light weight concrete blocks, autoclaved aerated concrete blocks, preformed foam concrete blocks, partial prefabricated concrete flooring and roofing units, concrete pipes, etc., all permitting use of fly ash and slag. Not only this, the Indian Standards on cement permit use of these alternative materials such as fly ash, slag, calcined clay, rice husk ash, etc. which help not only conserving our precious natural limestone reserves but also improve the durability of products and structures made these. Out of the 15 varieties of cements, for which Indian Standards have been developed, more than three-fourth of the cement produced in the country pertains to Portland pozzolana cement and Portland slag cement (popularly known in the market as blended cements). The Indian Standard Specification for masonry cement, intended for use in masonry constructions, permits use of various waste materials such as fly ash, calcined clay pozzolana, granulated slag, carbonated sludge, mine tailings, etc. Even, as per the Indian Standards, in the ordinary Portland cement, it is permitted to use up to 5% of these alternative materials, designated as performance improvers.

## Management of Steel Structural Work

Work related with civil foundations and all other subsoil work is a necessity of all projects. Thereafter comes the turn of structures either comprising RCC or Steel Sections. RCC structural work needs similar attention and considerations as the civil foundations. This book does not cover civil engineering practices and procedures and therefore we put to rest the matter pertaining to RCC structures.

However, if project comprises steel structures the first concern is procurement of steel structures of the required specifications that pertain to metallurgical requirements i.e., steel grade and composition and sectional properties.

Although this book is not meant for covering the subject of structural engineering (either theory or practice), I feel duty bound to give a piece of advice to project manager about what is, in general, desirable to be the method of assembly of structural members to form the complete structure as desired.

Methods of making connections between various fabricated members of a steel structure to enable desirable structure being finally given the desired shape. Connections are the glue that holds a steel structure together. Connections based of connecting medium are of three types, as below.

- Riveted connections (these are no longer used although almost of the old famous structures were constructed using riveted connections).
- Bolted connections,
- Welded connections and
- Bolted – welded connections.

Bolted, Welded and Bolted – welded connections are further briefly described below.

## Bolted connections

Modern practice with Bolted Connections is to use High Strength Friction Grip (HSFG) Bolts. The bolts are pre-tensioned against plates to be bolted together so that contact pressure, developed between the plates being joined, prevents relative slip when extra shear is applied.

Advantages of using HSFG bolting method of assembling steel structure are the following.

- Bolting operation is quite silent,
- Bolting is a cold process hence it does not involve any risk of fire,
- Bolting operation is quicker than welding,
- Manpower required is lesser but that of skilled category,
- Use of torque wrenches ensures that each bolt applies the desired contact pressure.

Disadvantages of using HSFG bolting method of assembling steel structure.

- If subjected to vibratory loads, such joints may result in reduction in strength and get loosened. This can be prevented by ensuring that the plates jointed by HSFG bolting are 'true' and do not have hard contact spots with each other and further their surface mating is no lesser than 85%. These conditions are near impossible to achieve in carrying out erection work at the site and therefore it is not advisable even to attempt to adopt them, if structure is supposed to be exposed to vibratory loads.

## Welded connections

There are three most common types of welds:

- Fillet weld (this type of weld is used most commonly

and is weaker than groove welds and other types of welds),
- Plug Welding (it is expensive and results in poor transmission of tensile forces),
- Slot welding (it is expensive and it also results in poor transmission of tensile forces) and
- Plug and slot welding (it is used to stitch, different parts of structural members, together).

Advantages of using welded connections.
- Using welded connections is economical in terms of costs of material and labour,
- Fabrication of complex structures is relatively easy,
- Provides rigid joints, which is modern practice.

Disadvantages of using welded connections.
- Welded process does not provide any provision for expansion/ contraction that occurs when molten weld filler material tries to glue with the metal of structural pieces by way cooling down and solidification. Therefore, chances of development of cracks in the weld seam are always very high and is can be imagined and understood a cracked welded seam is as bad as there being no real joint.
- This problem can theoretically be fully taken care of by carrying out thermal stress relieving operation at each significant weld. This, however, is practically impossible because it involves controlled heating and cooling duly maintaining a certain desirable rate of heating up to a certain high enough temperature (such as 700 C) and then holding that temperature for some time following by controlled cooling to ensure that there is no fast cooling. This process is adopted for some welded joints like those needed for fabrication a crane girder simply because it works out to be most cost-effective process for

the purpose.

- Uneven heating and cooling of structural members often results in distortion and may result in additional stresses.

- In a real-life scenario welding operations are required to be done in a large scale and it becomes impossible to inspect each and every welded joint, thus leaving quality of bulk of the welding job to chance – which is dangerous as cracks and distortions may in effect result in failure of structure under certain conditions of extreme stress for which the structure was designed and engineered.

## Bolted – Welded Connections
Advantages of using bolted-welded connections.

- Biggest advantage of this method is that most of connections are shop welded rather than being field welded thus making it possible to maintain high degree of quality control.

- Bolted-welded connections are highly cost effective as this method utilizes best practices of both the methods namely bolting and welding.

- Bolted-welded connections result in better effective strength and ductility of structure thereby enabling the structure to be really fit for bearing the stresses for which it is designed and engineered.

Disadvantages of using bolted-welded connections.

- So far as the genuine needs of project manager are concerned by way of ease and economy of carrying out assembly and erection of structures there is no disadvantage of bolted-welded connections.

- However, from the point of view of the persons responsible for furnishing necessary detailed drawing the detailing work is mainly brain stressing and needs well experienced engineering personnel as run-of-the mill drawings that can be furnished for Welded Connections would not serve the purpose.

- Therefore, engineering department or consulting engineer depending upon who is supposed to furnish assembly detail drawing in general try to evade and overlook the necessity of engineering the structure for bolted-welded connections and try to opt welded connections if they have their go.

The above is meant for narrating my personal experience and should in no way be construed to be suggesting firstly that engineering department or consulting engineers try to evade their duty to deliver their best and secondly that no matter what others advice and what has finally come out as the choice for assembling structural members, bolted or bolted-welded joints should be adopted just because it can be so inferred from the above. From this point of view, in fact, this entire book is gist of my experience in real life. Science is always dynamic and engineering puts scientific knowledge to advantage of mankind. Therefore, I would like to request reader to please bear in mind that whatever was "the latest technology" sometimes in the past is quite likely to have become obsolete today. Therefore, the reader is advised to try and visualize what can go wrong in the practice currently in vogue and take preventive steps well in time.

## Procuring of Steel Sections

There are a large number of steel manufacturers in India. An indicative list is given below, however, some of the manufacturers listed herein may go out of business and some new ones may appear on the scene.

| Name of manufacturer | Location |
| --- | --- |
| Shri Rathi Steel Dakshin Limited | Bhiwandi, Rajasthan |
| Shri Rathi Steel Limited | Gzahiabad, Uttar Pradesh |
| Jindal Steel and Power Limited | Raigarh, Chhatisgarh and Angul, Odisha |
| Tata Steel Plant | Jamshedpur, Jharkhand |
| Tata Steel Plant | Kalinganagar, Odisha |
| Visvesavaraya Iron and Steel Plant | Bhadravati, Karnataka |
| Bhilai Steel Plant | Chhattisgarh |
| Durgapur Steel Plant | Durgapur, West Bengal |
| Bokaro Steel Plant | Jharkhand |
| Chandrapur Ferro Alloy Plant | Chandrapur, Maharashtra |
| IISCO Steel Plant | Asansol, West Bengal |
| Salem Steel Plant | Salem, Tamil Nadu |
| Rourkela Steel Plant | Rourkela, Odisha |
| JSW Steel | Hospet, Bellary, Karnataka |
| Vizag Steel | Vishakhapatnam, Andhra Pradesh |
| Bhushan Steel Limited | Angul, Odisha |
| Rimjhim Ispat Limited | Kanpur, Uttar Pradesh |
| Essar Steel Limited | Hazira, Gujarat |

As has already been mentioned in connection with management of delivery of proper quality of rebar, malpractices also exist problems with respect to supply/ delivery of steel sections of intended and expected quality and it cannot and nothing should be taken for granted.

Unless the quantity of steel sections needed is large enough the buyer has no choice but to depend upon authorized distributors of various steel manufacturers for delivery of steel of the expected quality as ordered and in time. The authorized distributors try to push the product of the steel rolling mills who manufacture steel sections out of scrap steel, mainly ships with hardly enough control of metallurgy although the problem in case of large sections of structural steel is not as acute as with TMT rebar. Wide flanged beams are not manufactured by scrap steel re-rolling mills and therefore do not pose problems of quality, in general, although timely delivery is always a perpetual problem. For heavy steel sections such as Wide Flanged Beams and H Beams it is advisable to look into feasibility of importing exact cut lengths of such steel sections (a margin of total about 6 mm in length of each exact cut length may be provided for enabling end milling during fabrication) from countries such as Japan or China. For this purpose, project manager should take primary owner of project in confidence since payment of import duties and later obtaining refund thereof (in case of project location being a SEZ or upcoming Industrial Area notified by government) cannot, in general, be done by any party other than owner.

Subsequent to ensuring delivery of structural steel sections of proper quality comes the next step – fabrication and erection. My personal experience has been that fabrication at site is troublesome and needs too much and too stringent supervision else the finally fabricated structures are twisted, dis-shaped and of erroneous dimensions. Real problem starts with use of gas-cutting of steel sections with Oxy-Acetylene Torches rather than mechanical cutting, which cannot be done at site and even in fabrication shops mechanical cutting has its limitations with thickness of plates and their dimensions and therefore, use of Oxy-acetylene Torches has become quite common. Such cutting however, does result in rough edges that get overheated and in cases where edges of thick plates are to be used for butt welding, high-tech mechanical processes are made use of for clean cutting the edges for further fabrication by butt welding. Obviously, needed margins in dimensions are provided for mechanical cutting.

Good fabricators are needed to carry out necessary fabrication of steel structures in bulk. There are several steel structure fabricators in India. For example:

- Primax Building Systems Pvt. Ltd.. No. 496C, Sector-15. Faridabad – 121007, Haryana
- MALLAR GROUP, Plot No. 1&2, Peenya Industrial Estate, Peenya 1st Stage, Tumkur Road, Bangalore-560 058, Karnataka. India
- APEX BUILDERS, B-24 SECTOR-3, Noida, Uttar Pradesh
- Geeta Industries, 4/A, Narayan Textile Mill Compound, Dhan Bai Estate, Near C.T.M. Telephone Exchange, C.T.M. Cross Roads, Amraiwadi, Ahmedabad – 380025, Gujarat

How many of the above would still be existing at the time when you are reading this book, is anybody's guess. But some newer units would certainly have come up and finding out about them via Google or any other search engine should not be a problem. However, the golden rule of not taking words of any party on face value and maintaining continuous watch on progress at works of fabricator, irrespective of its repute, is the key to success. There will always be uncertainties in all actions and success would lie in successful bridging gaps caused by them. Although, it may be but obvious, I would like to mention here that while ordering structural fabrication care must be taken to ensure that all structural members are thoroughly cleaned of rust, scales and various surface defects by controlled sand blasting and then given two coats of primer paint at fabricator's workshop before being sent to project site for being used for erection. Another worth mentioning point, sometimes neglected at sites, is the proper storage of elements while waiting to be erected. Considering weather conditions, moreover in India due to its monsoons, elements have to be stored provided adequate conditions 'high and dry' in order to avoid the process of corrosion to be installed.

Structure can be conceived to be either finally assembled by welding or bolting or a combination of both, as has already explained earlier. However, irrespective of method of final assembly that is adopted, at this juncture I would like to say from my experience that key to successful structural fabrication lies is availability of accurate detail drawings. Generally detailing of fabrication is the job of engineering group or the consulting engineer responsible for designing and engineering structures. It is job of project manager to make sure that drawings pertaining to structural fabrication are made available to the fabricator well in time.

Moreover, someone – either the engineering department or the consulting engineer or the project manager has to be clear about sequence in which various members of the steel structure have to be erected. For this purpose the entire structure has to be codified in a scientific and logical manner so that members of structural erection team can decide and list out in advance various structural members that they would need handy at the erection site on any day for carrying out their job, organizing the storage of them accordingly.

There is no standard method of codification of structural members and each engineering group or consulting engineer adopts its own unique system of codification. Here I would like to mention one method that I have adopted. This method is described below.

Imagine a steel structure pertaining to some industry. More often than not, it is a lattice with rectangular base. The rectangular base is then comprised of rows of vertical structural columns connected by cross beams and diagonal bracings, the cross beams being at each floor, if the structure has to house a multistory factory or works. In a PLAN view the arrangement can be imagined to be comprising of a grid having several Rows and Columns. Generally Rows are lengthwise and Columns widthwise and Rows are identified by alphabets (A, B, C, D, ...) and Columns are identified by numerals (1, 2, 3, 4, ...) horizontally. If this system is followed various complete structural columns will be coded such A-1, A-2, A-3, ..., F-1, F-2, F-3. However, each column by itself may not be in one piece and it may comprise of 2, 3 or 4 pieces (or even more of them). Therefore, horizontally each floor of the building is given a number corresponding to its elevation above plinth level of the building. At plinth level obviously it would be "level-00", and if for example the first floor is a elevation of 4.5 meter above plinth level (or level–00) then it would be called "level-4.5". Next floor if it is 10 meters above plinth level will be called "level-10" and say the roof is at a height of 12 meters above "level-10" then it would be coded "level-22". Let us now consider the column A-1 and imagine that it comprises of three (3) pieces kept one above the other vertically, the bottom one being of a length of 8 meters, the next one again of a length of 8 meters and the top piece having a length of 5 meters. The bottom piece in this case would bear code "COL_A1/0-8" to signify that the piece belongs to a Column having PLAN matrix location "A1" and it would be installed between levels 0 meters and 8 meters. The piece above it would bear code "COL_A1/8-16" and the final piece would bear code "COL_A1/16-22". A member that is meant to be a bracing between bottom of Column A-1 and end of Column A-2 at 4.5 meters elevation, it would bear code "BRC_A1/0.0-A2/4.5". Similarly, all the structural members can be codified.

The fabrication detailing drawing also shows these codes, thus making it easier for the erection team to identify the pieces and to erect them properly.

Therefore, Project manager has to make sure that delivery of fabricated structural members is properly in sequence of their requirement at site for being erected because any structural member delivered to the site, out of sequence, will not only be useless but would also be likely to get lost and become untraceable when actually needed. At site a proper lay down area has to be identified for properly storing the fabricated structural members received from fabricator. For ease of picking up the required members from the lay down area it is necessary to also codify and mark rectangular spaces in lay down area. Doing so would go a long way in ensuring minimum loss of time in picking up the right structural members for the erection work planned for any time.

## Management of other erection activities inside structural building

Merit, of carrying out as many activities and much work as practically possible, in parallel, fully bearing in mind that the aspect of human safety and avoidance of damage to machinery, equipment or components, is well known to all experienced project managers because conversion of serial activities into parallel ones is the only tool available to a project manager for inserting floats in project schedule so as to provide for compensation of unfortunate and unforeseen time losses that are well known to be lurching around in industrial projects and kill progress because no matter how much carefully each and every activity and action is planned there does remain a chance for something or the other to go wrong. Thus, converting serial activities into parallel ones can be considered akin to saving for the rainy day.

From the above point of view, with all due precautions and thorough analysis to ensure safety while working and properly and carefully sequencing, planning and scheduling the work involved in various activities, the following jobs are being mentioned as examples only. Undertaking such activities and jobs in parallel with erection of structural building without waiting for completion of structural erection, has been found by me to be providers of time floats where there were none. Due care must, however, be taken to ensure that two different jobs do not go one simultaneously if one of them poses a risk of any material falling from a job area on to another job area that is directly below it. Few examples of such activities are as below.

- Erection of roofing of the building,
- Erection of prefabricated pipe supports,
- Erection of prefabricated cable tray supports,
- Construction of working platforms and intermediate

floors, if any, within the structural building,

- Erection of stairways and handrails.

It is not possible to produce here a comprehensive list of what erection work can be undertaken while the structural building is still under construction. The above are only indicative and project manager is advised not to jump into undertaking construction work inside structural building unless it has been assured and ascertained without any doubts that doing so shall not prove to be hazardous – resulting in bodily harm to any worker or damage to anything, however minor, that was supposed to be a part of the completed construction, irrespective of whether the thing that gets so damaged was a machinery or equipment or just a piece of structural steel or any other material. Once brain storming is done properly, it may be found that there are a lot of activities that can be undertaken simultaneously with erection of structural building. However, it goes without saying that doing so would not be a cakewalk and good enough safeguards would need to be made. It is advisable to use discretion and careful consideration including creative brainstorming among project team members before undertaking parallel activities under challenging circumstances. One usual way to conciliate erection works that in principal should not be carried out simultaneously due to safety matters is to organize different working hours / shifts or working areas in a way that both activities do not impact each other.

## Management of erection of tanks, vessels and pipes

Vessels and tanks to be erected in industrial projects are non-metallic as well as metallic. The non-metallic vessels and tanks are generally prefabricated and transported to project site duly complete and ready to be erected in their respective locations. Metallic vessels and tanks are often fabricated in situ, which is tricky and needs highly skilled workforce as well as sufficiently experienced engineer or technician to supervise the job. Under such a situation it is desirable that the engineer or technician concerned discusses the procedure and specific technical requirements well in advance of starting the actual fabrication. Doing this would enable to provide the engineer or technician good enough an idea about experience and competence of the workforce which in turn would enable him/her to look for replacement of those of the workforce who are not found competent enough or to arrange training them. Needless to say, that all required procedures must be strictly complied with and no compromise should be attempted. In case of non-metallic vessels and tanks due care and precautions are needed to be exercised to ensure that no damage occurs in the process of handling and erection and post erection all through various stages of progress of the project such care and precautions are needed to be strictly exercised.

In respect of prefabricated vessels and tanks due diligence needs to be exercised to make sure that the subcontractor or supplier of such vessels and tanks does not compromise quality in any manner while adhering to delivery schedule. All shop tests should be got witnessed and it should be made sure that instruments used for shop tests are properly calibrated.

Vessels and pipes are, more often than not, stored in lay down area for long enough time to allow internal rusting in vessels (unless constructed out of non-rusting material) and rusting as well as deposit of layers of fine dust inside pipes. Therefore, it is advisable to inspect vessels well in advance of due date of their erection so as to plan their cleaning either before carrying out erection or thereafter depending upon what is found to be most beneficial. However, so far as pipes are concerned their cleaning is a difficult process and calls for cleaning each piece individually. Real problem that arises is that unless special care is taken, the cleaned pipes get filled with dust and sand during handling by workmen. It is difficult to prescribe remedies for preventing this and different project managers are known to have tried different remedies for example, closing both ends of cleaned pipes by some means found most appropriate at site, like caps. However, my experience has been that best results are expected if pipes are cleaned in small batches rather then in bulk and bunches of cleaned pipes are secured and bundled properly and two ends of such bundle are covered properly for being transported to that portion of project site where they are to be erected and at erection site each pipe is flushed with dry oil-free compressed air. Compressed air flushing of pipes of up to 50 mm (2 inch) size is relatively easy. For bigger diameter pipes, dry oil-free compressed air needs to be used for blowing internal surface of pipes. Some site fabricated device made of pipe of suitable diameter and shaped to suit may be used at end of hose pipe used for blowing compressed air for cleaning internal surface of pipes of diameters larger than 50 mm and up to 100 mm. For pipes of diameter larger than 100 mm using pig made of gunny bags tied to manila rope may be used to scrap dust and dirt from internal surface of pipes. Cleaning of straight pipe lengths using above described methods is easy but cleaning pipe bends is tricky and project manager may hold collective brain storming sessions among his/her colleagues to device cleaning methods that may be most appropriate. The importance of this pre-erection cleaning procedure for pipes is factored for the ones dedicated for oil systems due to its strict cleanliness standards to be achieved during latter flushing.

While talking of vessels and pipes there also arises a question about whether or not any action is required in respect of various valves (shut-off valves and control valves). My experience has been that it is highly valuable not only to inspect and clean each and every valve but also carry out their pressure test so that no problem of any valve leaking or passing is faced during commissioning operations and thereafter. Once tested, the valves need to be handled with care and stored properly with adequate blanking of their ends so that ingress of dust and dirt and accidental damage to valve internals is ruled out.

With regards to timing of installation of valves at their intended locations in pipe work normal practice is not to install valves until all such activities that can result in damaging of valves are over. However, erecting pipes in absence of valves poses the challenge of providing gaps of exact dimensions in pipe work so that no problem is faced when it comes to installation of valves. Various practices are in vogue for this purpose. For flanged valves a good practice is to install flanges with their individual dummy blank flanges ensuring that dimensions of pipe work after removal of dummy blank flanges will be exactly as required and no problem would be faced in proper fitment of respective valves at the opportune time. In case of butt welded valves good practice is again to provide exact physical space in between the two pipes where the butt welded valve is to be located and then ensuring that the respective pipe ends are properly secured so as not to move; thereafter specially fabricated dummy bland flanges can be tack-welded using cleats so that the tacking is no lesser than 50 mm away from pipe ends where butt welded valve is supposed to be welded. Again, project manager would be benefited if a brain storming session if held and ways and means for ensuring that fitment of valves (flanged or butt welded) at the opportune time would be without any problems and until such time is reached the pipe ends are positively protected from ingress of dust and dirt or any other foreign materials. Some of pipes do have pump at one end and so far as pumps are concerned it is not advisable to undertake installation of any pump prematurely. Therefore, pipe erection work may be ended some meters away from pump with adequate protective measures at pipe ending so as to prevent ingress of dust, dirt and muck.

This book pertains to project management rather then being an erection manual for pipe work. All required details of technicalities for carrying out erection correctly are always provided in concerned engineering drawings. However, I do not consider mentioning a few rules of thumb to be out of place. These well-known rules of thumb in connection with erection of pipe work are as below, however, project manager is sincerely cautioned against taking decisions based on these without consulting engineering department or consulting engineer because there are several technicalities linked with each of these requirements that too must be taken fully care of and are beyond scope of this book. Some advices are given below and may be useful if such details are not given in erection drawings.

- Pipes for liquid medium should be laid with a gentle upwards slope (say 1 in 1000) in the direction of flow of liquid, so as to avoid air entrapment, reducing effective pipe cross-section and impairing flow. In case engineering drawings call for any bends in the pipe that are either vertically downwards or otherwise sloping down then provision of suitable vent with appropriate vent valves and tundish also need to be provided and these all have to be properly engineered rather being provided randomly without corresponding changes being incorporated in relevant drawings and other document. Engineering of these arrangements is intricated and due consideration is to be given to the liquid that will flow in each particular pipe work. What is required for water would not be appropriate for oils or any aromatic fluid.

- Each steam pipe needs also to be first laid in a gentle downwards slope (say 1 in 1000), in the direction of flow up to some distance and then the direction of slope is reversed to become upwards in the direction of flow. The purpose of doing this is to create a location in the pipe work, where direction of slope changes from downwards to upwards that serves to act as a collection point for the condensate (which is always produced whenever steam is passed through pipes that are cold). At this lowest point where direction of slope of steam pipe changes from downwards to upwards steam-trap with well engineered root valves and tundish arrangement need to be provided. Location of the point where the direction of slope of steam pipe is to change would depend on overall length or run of the steam pipe without any bends in 'Z-direction' (horizontal plane is considered in 'X-Y' direction and vertical in 'Z' direction). At each location where change in direction of slope changes from downwards to upwards will need similar arrangement of steam-trap and tundish. However, necessity of entire arrangement has got to be properly engineered and incorporated in drawings before actual physical execution.

- Taking care of phenomenon of water hammer is another important matter of intricacy of engineering of piping systems and it can be safely presumed that Engineering or Consulting Engineer whosoever has engineered a piping system has taken care of technicalities of the phenomenon. The subject in itself is quite complex and nothing much can be prescribed as a magic remedy for all kind of pipe work. It is expected that project manager is well informed and familiar about the phenomenon. With this in view the subject matter of water hammer is being described below in as simple a language and manner as possible.

  The phenomenon of water hammer, as the name suggests may be mistakenly assumed to be typical to pipe work meant for carrying water. However, the fact is that it is equally applicable to all pipes carrying any liquid. Basically, the phenomenon arises because of (a) property of incompressibility (in reality each liquid can be very slightly though practically negligibly compressible) of liquids and (b) law of conservation of energy in nature.

Pipelines, in general, are engineered assuming that velocity of liquid in any pipe is generally constant and changes of velocity, if any, when they occur for controlling flow of the liquid are slow enough and smooth rather than being abrupt. In real life this assumption generally holds good since flow control valves have not only a finite operating time when opening to increase or closing to reduce flow of the fluid but also the rate at which they cause the flow velocity to change is quite slow.

Liquid flowing inside a pipe can be imagined to be a cylindrical mass traveling inside the pipe. Newtonian physics tells us that mass tends to remain in its state of rest or uniform rectilinear motion unless a force is impressed upon it and that when a force is applied it tends to accelerate the mass such the acceleration is proportional to the applied force, proportionality constant being mass itself. A person is hurt if someone throws a stone aimed at him/her because upon being thrown the stone is at a certain velocity and when the stone strikes a person its velocity is suddenly (it being sudden is in reality a concept only; what really happens is that stone first touches skin and then skin resists its motion and in fraction of a second stone velocity is brought to zero) brought to zero which requires enormous force to come into play to decelerate the stone to zero velocity. It is really this decelerating force that causes hurt. Thus, basically any transition in velocity of flow of fluid in any pipe calls for forces coming into picture.

In real life all applications of pipes and valves face transitory behaviour of flow of liquid – for example opening a valve causes liquid velocity to rise from zero to certain other value and closing of valve similarly results in fluid velocity to come to zero from a certain other value. The slower the rate of changes in velocity of flow in pipe the lesser the forces that come into play. A sudden change behaves in the same manner as a stone being hurled at something or someone and therefore bring into picture corresponding accelerating or decelerating forces. Moreover, pumps that supply fluid into pipes may have a Q/h characteristics that may also result a rise in static pressure of fluid in pipe when flow of liquid in brought to zero by closing a valve, irrespective of whether such closing of valve is slow or not.

The worst situations arise when flow of fluid in a pipe is by design needed to be abruptly (that is in time as close to zero as practically feasible, say typically in about 50 to 200 milliseconds depending upon application) reduced to Zero meaning thereby that flow from being say Q liters per second is very fast brought to Zero liters per second. This occurs when 'Slam-shut Valve' is used for practically sudden stopping of flow of liquid in any pipe. There are many applications in which Slam-shut Valves are used. A typical example can be the case of a gas turbine using a liquid fuel and used for generating electricity by way of driving a generator. In case of occurrence of any electrical fault in system it is required that the Circuit Breaker (CB) connecting the generator to electrical system is "TRIPPED". Typically, a CB takes 10 to 60 milliseconds to operate and disconnect generator from system. Obviously, therefore, it becomes necessary, in this instance, to engineer the fuel supply system of the gas turbine to be equipped with 'Slam-shut Valve' to meet this exigency.

I faced a real-life situation once as follows. The matter pertained to a Naphtha fuel based combined cycle power plant comprising a Gas Turbine Generator (GTG) of Max. Continuous Rating (MCR) of 44 MW exhausting flue gases into a heat recovery steam generator with its Steam Turbine Generator (STG) of MCR of 22 MW. Naphtha being a highly volatile and flammable liquid the entire Naphtha handling system, including sliding roof storage tanks and

Naphtha pumps for supplying Naphtha fuel to gas turbine generator, was so engineered as to be distant and secluded enough from any habitat and was at a distance of about 2.5 km from the GTG.

Owner of the power plant was also owner of the power plant engineering group which was basically engaged in an 'Remote Detailing from Engineering of Structures' job on contract and was handling back-office engineering and detailing work for a top-level consulting engineers firm in USA. However, the said power plant engineering group had never carried any power plant engineering work independently. Moreover, detailing is simple but engineering is complex therefore doing the engineering it is not a cake walk, for those who are used to detailing only. My job responsibility was to oversee the engineering work pertaining to the said Naphtha fuel based power plant and to provide guidance to the engineering group in cases where the engineering carried out by it was either erroneous or otherwise doubtful and I was supposed to obtain expert inputs to the extent needed from top grade experts of relevant subject so that the engineers pertaining to the power plant engineering group stood appropriately corrected and the power plant engineering group was supposed to take a learning for being ready for handling such responsibility independently in future. Obviously, my job was tough and I had to look into the engineering, so to say with magnifying glass. Several minor as well as major mistakes came to my notice, but relevant to the topic of piping is the following.

When I received details of engineering of Naphtha forwarding system I noticed the following:

- Naphtha forwarding pumps with nearly flat or horizontal Q-h characteristics were conceived so that there may not be any serious rise in static pressure of Naphtha in the 2.5 km long delivery pipeline in case valves at turbine end are closed.

- For delivering Naphtha to gas turbine at a rate

of 13.05 tonnes/hour corresponding to MCR rating of the gas turbine-generator, pipeline comprising seamless carbon steel pipes of grade ASTM A106 Grade-B, schedule-60, having size DN100 (100 mm ID) was provided. A Slam-shut Valve and a gate valve were provided at end of the naphtha delivery pipeline and piping from the gate valve was taken to inlet point of the gas turbine. The Slam-shut Valve had a closing time of 50 millisecond and was hooked to the GTG controls so that along with a tripping command to the GTG a Closing command would also be applied to the Slam-shut Valve.

- Upon going through engineering details of the Naphtha forwarding system I found that the possibility of occurrence of water hammer in the Naphtha delivery pipeline was overlooked. Therefore, I performed rough calculations of pressure that would develop in Naphtha delivery pipeline at delivery end close to the Slam-shut Valve if it closes suddenly (i.e., nearly in zero time) when Naphtha is flowing in the pipeline corresponding to MCR demand of the GTG (detailed computations involve mathematical modeling of the system duly considering that a closing time of 50 milliseconds is not same as zero time and flow rate in pipe doesn't abruptly drop to zero in no time but gradually reduces to zero depending upon valve characteristics – also due consideration is made of static pressure wave traveling in the concerned pipeline). The rough calculations were as tabulated below.

| GAS-TURBINE GENERATOR & NAPHTHA DATA: | | |
|---|---|---|
| Density of Naphtha Fuel at 15C | kg/cum | 665 |
| CV of Naphtha Fuel | kJ/kg | 43932 |
| Gas turbine-generator (GTG) Rated Output | kW | 44000 |
| Heat Rate of Gas turbine-generator | kJ/kWh | 10741 |

| (GTG) | | |
|---|---|---|
| Input Energy into GTG | kJ/hr | 472604000 |
| Corresponding Input of Naphtha Fuel | kg/hr | 10,757.63 |
| | Cum/hr | 16.18 |
| | Cum/Sec | 0.0045 |
| **TRANSIENT FORCES CALCULATION:** | | |
| Naphtha Transport Pipe Size (ID) | meter | 0.1 |
| Pipe Cross-sectional Area | Sqm | 0.0079 |
| Flow velocity of Naphtha in pipe | m/sec | 0.5721 |
| Pipe Line Length | meters | 2500 |
| Pipe ID | meters | 0.1 |
| Pipe Cross Section | Sq m | 0.0079 |
| Internal Volume of Pipe | Cum | 19.6350 |
| Density of Naphtha | kg/Cum | 665 |
| Mass of Naphtha inside Pipe | kg | 13,057.2445 |
| | tonnes | 13.0572 |
| Volume of Naphtha flowing in pipe at full load | Cum/Sec | 0.0045 |
| Velocity of flow of Naphtha in pipe | m/sec | 0.5721 |
| Time for slam shut valve to Close | milli sec | 50 |
| | Second | 0.0500 |
| Naphtha Flow Deceleration on Closing of Slam Shut Valve | m/sec/sec | 11.4428 |
| Pressure corresponding to Deceleration (Force = Mass x Acceleration) | kg(force) | 149,411.4642 |
| This force can be imagined to be applied through a piston of same | kg/Sqcm | 1,902.3658 |

| diameter as that of Naphtha transport pipe, hence corresponding pressure | | |
|---|---|---|

As can be seen the value of surge pressure at delivery end of Naphtha delivery pipeline had worked out to be large enough to result in bursting of the pipeline.

This was alarming. I advised the person in-charge of the engineering group, informing him in full details about my apprehension and advised him to try and obtain second opinion from the American Consulting Engineering firm about my apprehensions. The said American firm seconded my opinion. Subsequently, mechanical engineering department of a reputed engineering college was consulted and after carrying out detailed analysis using computerized simulation technique as well as mathematical modeling the engineering college suggested provision of a return pipeline of size 40 mm, tapped from the 100 mm Naphtha delivery pipeline just before the Slam-shut Valve, run along with the Naphtha delivery pipeline at a gentle slope of 1 in 5000 rising up from near Slam-shut Valve up to the Naphtha tank and connected to suction header of the Naphtha Forwarding Pumps as also to top of Naphtha storage tank and with the provision that the root valve at tapping point from Naphtha delivery pipeline just before the Slam-shut Valve would in general always be kept in 'Open' position whenever the GTG is operated. The studies carried out by the Engineering College showed that with this arrangement the pressure rise in the Naphtha delivery pipeline upon closing of Slam-shut Valve when the GTG is at Full Load (MCR) will be only 10% of normal working pressure.

The reason for citing an experience in the above is to bring home the point that keeping eyes and ears open and looking for possible errors rather than taking for granted that the engineering will always be right can sometimes be helpful, in warding off occurrence of serious mishap. However, seldom do the project managers have the luxury of sufficient time float so as to be able to indulge in trying to find fault in something that normally is supposed to be flawless, but if luck or chance does present a situation of availability of some good enough time float, looking into engineering drawings and calculations with magnifying lens may prove worthy of the effort.

## *Management of Installation of various Equipment, Plant & Machinery*

One basic fact about all industrial projects is that for each and every material thing, equipment, plant and machinery there is an opportune time, basically depending upon completion and readiness of preceding activities, before which it is counterproductive to undertake installation of that thing. Therefore, it is highly advisable that start of installation of equipment, plant and machinery is not started prematurely lest outcome may be disastrous and may result in cost and time extension rather than reduction.

In real life it is not possible for any project manager to personally supervise each and every activity pertaining to any sizable industrial project. Good practice is that managers are deployed for management of various project activities – for example Civil Engineering Manager, Electrical Engineering Manager, Chemical Engineering Manager, Controls and Instrumentation Manager, IT Equipment and Implementation Manager etc. What is required of the project manager is that he/she does make sure that the unlikely and undesirable instances, of any of these respective managers taking chances or compromising the specific requirements of installation of equipment, as required by good engineering practice and prescribed in Erection Manuals provided by respective OEM.

Erection and handling machinery such as cranes, winches, hoists, come along tools, chain pulley blocks, crowbars, wooden sleepers, steel plates, steel girders, wire ropes, hessian ropes etc. are used apart from various hand tools and mechanized handling tools are used for carrying out the work of installation. If due care in meticulously taken to always ensure that each and every machinery, implement, tool and material that is used for carrying out some installation work is free of flaw such as fatigue, cracks, insufficiency of capacity to handle forces there is no reason why there should be any occurrence of undesirable falling, toppling and damage of any equipment and/or loss of life.

It is desirable that proper schedules are drawn out for checking, testing and inspection of all that is used for carrying out installation and that no compromises are ever made. In real life, however, people tend to take things for granted and believe word of mouth rather than making sure by way of checking schedule document to ensure that nothing is lacking. Each instance of taking something for granted whereas a flaw, however minor does exist, is like a straw loaded on a camel's back. Addition of such straws one by one constitutes load and ultimately there comes a stage that all that is needed is addition of one last straw on camel's back that breaks the back causing the camel to collapse.

## *Special features of installation of Control & Relay Panels and Cables.*

For start of installation of electrical part of a project it is necessary that civil work is in condition of reasonable completion else there can arise a highly undesirable situation resulting in ingress of dust, dirt and moisture in various components installed inside various panels and rendering electrical cables partially damaged due to twisting, totally undesirable short radius bending and application of cable pulling forces on conductor cores.

In general, none of the following civil work should be going on concurrently within a radius of about 50 meters when installation of electrical equipment etc. is carried out:

- Excavation and/or earth filling,
- Consolidation of soil,
- Road construction,
- Plying of vehicles on roads that have not been Asphalt Paved,
- Cement-sand plastering of walls of room in which electrical equipment has to be installed,
- Fixing of doors and/or windows or window blinds in the room in which electrical equipment has to be installed,
- Fixing of concrete anchors (for any purposes whatsoever) on the roof in the room in which electrical equipment has to be installed,
- Wall painting of room in which electrical equipment has to be installed,
- Installation of false roofing in the room in which electrical equipment has to be installed,
- Installation of air conditioning ducts or the like in the

room in which electrical equipment has to be installed,

- Any other construction or installation work that involves persons other than electricians and/or electrical supervisors and engineers in the room in which electrical equipment has to be installed (reason being such personnel are generally not aware and cautious enough to make sure that they are not bringing in some dust either with their foot wear or clothing),
- Any such other work that may cause ingress of dust and dirt or chemical fumes in the room in the room in which electrical equipment has to be installed.

Main purpose of listing the above is that more often than not what really happens in most of industrial project is exactly the opposite and being under pressure of shortage of time project managers often find it unavoidable to ignore the above fundamental requirements of ensuring that electrical equipment meant for installation in a project doesn't suffer from what can be considered as birth defects – the defects that make the equipment installed in industry to become prone to error and failure right from day one. While speaking about electrical equipment, it is desirable that we look into nuances of installation of each type of electrical equipment that constitute the science and art of project execution.

## Installation of cable trays and cable carrying conduits.

Unless arrangements for laying cables are properly in place there can't be any meaning or purpose of installing other electrical equipment unless physical situation at site demands doing so – for example if rooms, where electrical control panels and/or cabinets have to be finally installed, are ready complete in all respects so that respective equipment if installed in their right locations in a final manner ensures that such equipment would be safer in this manner even though laying and connecting of cables meant for them may have to wait for some time.

Good electrical engineering, in connection with industrial project calls for making proper Cable Schedules available to project execution team. Proper Cable Schedules have the following features.

- There is a scientifically worked out Alfa-numeric Nomenclature of each and every individual cable that would be laid during execution of project. Scientifically worked out nomenclature includes short names of equipment at either end of cable, core conductor material (AL/CU), number of cores in cable, core conductor cross section, type of cable (armoured/unarmoured or shielded/unshielded or with/without ground wire (earth wire), insulation material, sheath materials.

- There is a drawing or chart showing against each cable (i.e., against the Alfa-numeric Nomenclature of each cable), cut length of cable needed for laying from end to end and in case the cable is laid inside conduits (or ducts), the names (Alpha-numeric) of conduits or ducts it has to pass through, other cables (their Alfa-numeric Nomenclature) that have to be bunched together with it when being laid inside conduit or duct.

Cables should be handled with care making sure that they

would each be reeled off from their respective cable reel so that no cable is twisted. If it becomes unavoidable to take any cable out of its reel and placing it on ground then it should be kept on ground in shape of '8' forming one shape of '8' above the other. Depending upon cable diameter, physical dimensions of '8' can be anything between half a 0.5x1 meter to 3x6 meter or as needed.

Cables can't be put inside any conduit of significant length by pushing it from one end. Rather, cables have to be pulled through conduits with help of a pilot rope that has already been laid inside full length of the conduit concerned. Obviously each conduit needs to be having a pilot rope laid inside it. Also the pilot rope has to be strong enough so as to be capable of successfully bearing the force of pulling.

For pulling cables conduits the first requirement is that a strong and flexible enough rope (generally nylon rope or wire rope if really unavoidable) is passed through each individual conduit or duct so that the cable (or bunch of cables – as the case may be) can to tied to this rope at one end and the rope can be pulled from the other end. This is best done right at the time of installation of conduits by laying a thin nylon rope (of size 6 mm) through each conduit with help of G.I. wires of 12 or 10 SWG as pilot. In case it is needed to pass wire rope through any conduit then first nylon ropes of increasing size should be laid and then finally when a thick enough nylon rope has been laid it should be used to pull the wire rope of desired size. Pulling of the guide

For pulling cables through conduits care should be exercised to make sure that pulling force, rather than being applied on to conductors of cables should be applied on to cable sheath. Specially designed Cable Pulling Grips should be used for this purpose. One of the good Cable Pulling Grips is "Rotating Eye, K Type Pulling Grip", made of high strength galvanized steel strand. These are to be specially imported from USA because in India there is hardly any awareness about it and the procedure normally used for cable pulling is detrimental to life of cable and because there is almost no demand of such grips these are neither manufactured in India nor readily available with suppliers of tools and tackles. All K Type Grips feature double weave mesh for greater strength and added mesh contact on the cable to handle longer or heavier pulling jobs. Best part is that if such grips are used for pulling cables, no pull is applied on any single part of cables and the grip, in effect, behaves as if the cable is held by hands and pulled – thus ensuring that cables sustain no damage due to axial pull applied. The K Type Pulling Grips are equipped with forged eye that mates easily with a swivel or shackle. They are specially designed for use in the installation of underground power cables, communication lines and service lines into factories, shopping centers, construction projects and general underground electrical constructions. (For further details reader is advised to search for "K Type Cable Pulling Grips" on Internet. It is advisable that "Cable Pulling Grips" of all types and make should be looked into for their merits and specialties.

However, as time passes such grips may become commonplace and may be manufactured in India or at the least stocked and sold by tools and tackles suppliers and Mill Stores in India. Therefore it may be useful to try and look for indigenous availability of these grips in India before hastily importing them, as there may be substantial price difference.

## Installation of electrical equipment, plant and machinery.

This book pertains to management of project rather than the procedure for installation of Electrical equipment, plant and machinery. Therefore, skipping the procedural of installation we come to the subject matter that is specific to installation with specific emphasis upon the matters of utmost importance in relation to installation part that forms the core of execution of project. Electrical equipment, plant and machinery always comprise of electrical insulation and dielectric materials in various forms – solid, liquid and gaseous. Electrical properties of these materials are subject to deterioration in presence of contaminants such as moisture (including the invisible atmospheric moisture), chemical fumes, dust and dirt. Electrical equipment, plant and machinery also more often than no have magnetic core made of steel that can rust and corrode in presence of atmospheric moisture and chemical fumes. Therefore, it is advisable to take special meticulous care to ensure that all electrical plant and machinery remains free from ingress and attack of any undesirable contaminants such as moisture, chemical fumes, dust and dirt.

It is not feasible to give here good enough guidelines in this connection because requirements would vary on a case-by-case basis and there is no cure all single remedy. Mentioning here any one or two specific procedure is more likely than not to create confusion. The best practice is to look into specifics for each and every individual electrical equipment, plant and machinery. Generally, OEMs do provide good enough Erection Manuals mentioning therein all requirements specific to each equipment, plant and machinery.

What happens, more often than not, is that post installation of electrical equipment their commissioning and energisation occurs after long time gap. Time gap of more than two weeks is long enough for transformers, motors, generators, circuit breakers as well as control and relay panels to result in start of deterioration of respective equipment and calls for undertaking their regular maintenance. Ground reality in all industrial projects, in general, is that this requirement of regular maintenance of electrical equipment is almost never factored in planning of execution of projects. As a result, no engineering set-up is available to project manager for ensuring the required proper maintenance. Maintenance is a different cup of tea and those who are expert in project execution may have no idea of what is required and how to manage this. On top of this necessary specialized workers and supervisors for carrying out the required maintenance work are not available at project site. Maintenance of equipment requires not only carrying out certain work but also maintaining records pertaining to periodic tests and maintenance work done. Just for example, an oil filled transformer needs (i) testing of Insulation Resistance and Polarization Index of various windings, (ii) testing of insulating oil for dielectric strength, (iii) regeneration of silica gel dehydrant of breather of conservator tank, (iv) inspection to ensure that no leakage of insulation oil is occurring, (v) cleaning dust and dirt from surfaces of electrical bushings as well as transformer tank (vi) testing of insulation resistance of cooling fan and insulation oil pump motors, if present. The job doesn't end with the above. What is required is properly maintaining records of all the above and other work carried out and ensuring that these various jobs are repeated at intervals suitable for each of them. In case any tests indicate start of deterioration of any sort immediate action to mitigate it need to be taken, under intimation to and supervision of respective OEM's engineer. Also, it is required that all the records of maintenance work are handed over to project owner (maintenance and operation department in case a project is being executed in-house) and this handing over of records is properly documented.

The above has been described just as an example. Detailing maintenance procedure for all types of electrical equipment, plant and machinery here will be out of context in this book at the very subject matter hereof is management of project rather than maintenance of equipment.

## Management of installation of Mechanical Equipment, Plant and Machinery

Regarding installation of mechanical equipment, plant and machinery also we would not touch the step-by-step procedures to be followed for installation. All equipment need taking care of precision and handling without damaging what is under installation. It goes without saying that normal precautions would be taken in usage of cranes and other installation machinery.

If normal as well as equipment specific installation procedures given in installation manuals supplied by Original Equipment Manufacturer (OEM) are adhered to there should be no reason for occurrence of any problem whatsoever.

As has been said in the above regarding electrical equipment, plant and machinery; mechanical equipment, plant and machinery to is prone to start deteriorating if left unused and therefore unattended for long enough a time, post their respective installation. For example, all machinery having journal bearing need regularly being supplied with a preserving lubrication oil for preventing journals getting rusted and same way the machinery using anti-friction bearing too need care for prevention of balls or rollers, races or cup and cones and ball/roller retainer rings getting rusted. Same way gearboxes are prone to damage to their bearings and/or journals and gears in the long run if left unattended for long. It is not possible to provide here a tabulation of all kinds of equipment and what sort of deterioration can occur in each of them on being left unattended after installation. OEMs can provide details of equipment specific preservation methods and it is highly desirable that project manager consults the respective equipment OEM for obtaining details of preservation methods and then applies the preservation. It is equally important to maintain a Log Book for recording, in details, all the preservation procedures from beginning to end. Such Log Books come handy should any dispute arise at the time of handing over project to owner in respect of deterioration having occurred due to time gap between completion of installation and start of commissioning operations.

## *Management of Installation of Intra-Plant Communication (Intercom) Systems*

The Intercom Systems should be installed as the last one when all construction and heavy erection and installation work has been completed. Typically, the following should have been completed.

- All work pertaining to civil and structural erection including its painting.
- All piping-work including painting and cladding, if any.
- All electrical cable laying work.
- Permanent lighting and ventilation systems should not only have been installed but also finally commissioned – there should be no temporary flood lights any where in use of kept in the plant building for any purpose.
- All temporary staging and scaffolding should have been removed and permanent access arrangement for each piece of instrument should have been completed in all respects.

## *Installation of Instrumentation & Controls (I&C) Equipment.*

Owing to their delicacy I&C Equipment should be the last ones to be brought to their final locations. The following should be the status of readiness of the industrial plant building for being considered to be fit for finally locating the I&C Equipment.

- Compressed air system, if meant to be used for I&C

should have been commissioned and compressed air should duly have been tested for required dryness and freedom from lubricants.

- In case the I&C System has been conceived based of Distributed Processing with endless Optic Fiber Cables for connection to central computer for controlling individual control centres then proper sealed (suitably cooled) dust and dirt free cabinets for housing various Transmitters, Signal Conditioners (A/D and D/A converters), PLCs etc. should have been installed with their respective power supply available duly tested for being as required, cooling system duly tested and fiber optic cable duly laid and ready to get connected to the respective I&C Equipment meant to be housed in all of the cabinets.

- Each of the I&C Equipment should have been tested and calibrated in test lab so that upon being installed in respective cabinets in the field in industrial building or outside near any outdoor equipment that needs to be controlled, it can be relied upon for performing during various commissioning test without throwing surprises.

## *Installation of other equipment that may be specific to any project.*

Technology is so fast advancing that it is not possible to take up and discuss here the precautions and prerequisites of management of installation of each and every type of equipment that may be required to be installed in all various types of projects that a project manager may be required to handle.

Nonetheless, some simple rules for handling Installation of equipment irrespective of its type and purpose precipitate, as below, from what has been discussed above regarding certain type of commonly encountered type of equipment in all projects.

- For each type of equipment there is requirement of a certain level of completion of other project activities before which it would be premature to undertake installation of that particular equipment. It is advisable that in case of lack of clarity about this aspect a project manager should better consult the OEM concerned of that equipment.

- There may be instances when a project manager himself is not very clear about specific requirements of installations of any particular equipment and no really experienced expert is specifically available for undertaking that work based upon previous experience. Under such circumstances it would be foolhardy to venture into a daring attempt to start such installation. It is highly advisable that OEM be consulted and detailed guidelines be obtained for the work. Better still would be availing services of Erector or Installation Technician or Installation Engineer of the OEM, if not for carrying out the installation at least for supervising it provided qualified manpower is available with the project manager at the site for carrying out the work under supervision of OEM's person. Care should be taken to ensure that there does not arise a situation where meaningful dialog or conversation may not be possible between the OEM's person and the project site manpower. For example, if installation of a robotic machinery made in China is to be undertaken, there may be a serious problem of Indian manpower available at site to be able to make out what the OEM's person is advising. It is not possible to suggest here the real remedy under such circumstances but, in nutshell it may be useful to get installation carried out by the OEM using their manpower (for which there may arise problem of arranging India VISA for OEM's manpower). Normally, such issues are addressed in contracts and procedures are quite clear but in a rare case if this aspect is overlooked in the beginning there is every like-

The Science & Art of Management & Execution of Industrial Projects in

lihood of cost as well as time overrun.

## Management of Commissioning of Industrial Project

Generally, the work of commissioning is handled by a team of commissioning experts who often are too high headed and consider a project manager as no more than a facilitator and provider of facilities and materials. However, ultimate responsibility of ensuring timely commissioning of project rests with project manager. Therefore, it is important that project manager is able to certainly but politely and tactfully assert himself/herself and is able to (i) continue to be the hub of communication, (ii) ensure that there is good rapport among various experts, (iii) identify roadblocks and manage to remove them and (iv) make out that time is being wasted in undesirable trial-and-error type of handling during commissioning operations in cases where equipment or plant seemingly appears to be all okay but still some unknown phenomenon interferes with achievement of reliable operation and desirable end result.

My experience has been that in many of such cases the culprit, more often than not, is some 'black-box' equipment (Black-Box equipment or apparatus are those that are complete in themselves and do not allow anyone to look at them internally, for example a PLC) happens to be the culprit and it pays to just replace such black-box equipment or apparatus with a new one. However, if such replacement of the block-box equipment appears to be helpful, it is further advisable to once more replace it with another new piece by way of a confirmatory test that the particular black-box equipment was misbehaving. Such a situation becomes difficult to handle in cases where spare new black-box equipment is not available. Such cases are really unfortunate because project manager just cannot afford to quickly procure a spare piece of black-box equipment just for making a trial out of his/her hunch to see whether or not replacing the suspected black-box equipment helps simply because if such a trial proves to be unsuccessful the expense in procuring the spare becomes unjustifiable. Handling such situation is quite tricky as precious time goes on ticking and deadline of completion of commissioning continuously keeps approaching, giving project manager sleepless nights. If such a situation does arise it may be best action on part of project manager to take his/her boss in confidence and explain to him/her about the situation so that an expert may be deputed to project site for solving the problem.

An industrial project, so far as technical aspect is concerned, is generally a combination of a few different plants. For example, a Water Demineralizing Plant which by itself is a complete plant and in fact can even be a big enough industrial project in a scenario where Government promotes some Special Economic Zone (SEZ) in a coastal area due to which almost all the industries coming up there can be likely to need demineralized water and some entrepreneur may invest in setting up a sizable water demineralizing plant. Whereas, if the project concerned pertains to setting up of a thermal power plant it would invariably be planned with an integral water demineralizing plant irrespective of there already existing a sizable commercial Water Demineralizing Plant as an independent industry. This may happen because modern super-critical pressure boilers, used in large size thermal power plants need extremely pure water for which supply from any outside source is not desirable.

Therefore, each complete plant, within a project that can technically be considered as a small project needs to be commissioned as an individual entity. Project manager has to so plan that as many as possible out of these plants comprised in the project at hand, are commissioned individually either in parallel or in certain sequence, as may be specifically required.

Other than individual plants there are systems in each project that need to be commissioned. For example,

- Electrical system, which may comprise of some arrangement of receiving power from a source or power supply company. There may be several levels of voltages, for example Extra High Voltage (EHV such as 400kV or 220 kV or 132 kV or 66 kV or 33 kV), High Voltage (HV such as 11 kV or 6.6 kV), Medium Voltage (MV such as 400 Volts) and Low Voltage (such as 110 Volts) and DC power system comprising storage batteries and battery chargers;

- Fuel receiving and delivery system;
- Chemical receiving, storage and delivery system;
- Raw material handling system;
- Firefighting system;
- Waste storage, handling and disposal system and
- Various project specific systems.

Some of the plants and systems of each industrial project are such that their commissioning can be undertaken simultaneously whereas some plants and systems are such that their full commissioning cannot be undertaken unless a certain other plant or system has been commissioned and is ready to perform a certain requirement.

It is but obvious that project manager has to regularly hold joint meetings with the various commissioning managers or experts who have to undertake the responsibility of commissioning various plants and systems and coordinate their activities.

Commissioning aims at driving a project towards its ultimate completion and bringing it to a stage where it is able to be put to commercial use for production of the intended entity. Therefore, commissioning sooner or later becomes akin to actual commercial operation. Quite obviously, therefore, it has to be the first priority for a project manager to put in place a foolproof system of issue of permits for doing work on various plant and machineries. Also, manpower of quite a different type has to be deployed for safely undertaking commissioning activities (basically Commissioning Operation). From the moment of start of commissioning it does not take long before plant and machinery, if not handled properly and safeguarded from being approached by anyone who is not authorized to do so, can pose serious threats of injuries, including fatal, to persons and damage, including very serious damage, to machinery. Therefore, all safeguards too have to be in place including Work Permit System and Safety Locking System (Lockout-Tagout) as if the project is already in operation and production.

Commissioning of all types of industrial projects comprises of the following activities, in general.

- **Mechanical Completion of installation (including that of electrical equipment):** Mechanical completion means that all the plant and machinery that was to be erected, installed and interconnected is actually physically in place, duly erected as required, without any exception. It is a universal practice that the manager concerned in-charge of mechanical completion of various different plants and sub-systems certify mechanical completion of the part of project falling within their respective jurisdiction. Mechanical Completion Certificates (MCCs) constitute a very important part of documentation of project.

- Checking of all Piping & Instrumentation Diagrams (P&ID) and comprises detailed diagrams which show the piping and vessels in the process flow, together with the instrumentation and control devices).

    Checking of P&IDs at the juncture of commissioning is not meant for checking and ascertaining correctness of the P&IDs themselves – rather what it really means is cross checking of entire installation to make sure that there remains no discrepancy between actual execution and the diagrams.

- **Pre-Commissioning (PC) checks:** The Pre-commissioning (PC) is really a construction activity that involves the verification of functional operability of various elements within the system to achieve a state of readiness for the Commissioning and Start-up operation. Sometimes, for the pre-commissioning of Process units requires the completion of the commissioning activities of utility facilities like power supply system, water supply system etc. The Pre-commissioning may require an energisation of equipment includes the running and entire testing of the power generation systems, testing of emergency shutdown and control systems and dynamic trials prior to commissioning and start-up operations. Sometimes it becomes necessary that the Pre-Commissioning activities be completed prior to the achievement of the Mechanical Completion Certification (MCC).

The pre-commissioning tests need to be carefully programmed so that they take place in a logical and efficient order, in order that no equipment is disturbed again during subsequent tests

It is not intended to list down all the possible Pre-commissioning checks or tests for all kinds of industries or systems. However, just in order to give a clear-cut idea of what the Pre-commissioning checks or tests comprise of, a list of the requirements for Pre-commissioning of Electrical Switchgear and Motor Control Centers is being given below. In addition, the list of the tests to be carried out should be arranged in a chronological order together with any precautions that need to be taken into account.

- Analysis of the wiring diagrams to confirm the polarity of connections, positive and negative-sequence rotation, etc. of rotating equipment,

- A general inspection of the equipment, physically verifying all the connections, at both the relay and panel terminations,

- Measurement of the insulation resistance of cables and equipment,

- Inspection and secondary injection testing of the relays,

- Testing current transformers including checking their polarity and primary injection testing,

- Checking the operation of the protection tripping and alarm circuits.

- Dry Commissioning of plant and machinery: In Dry commissioning, various tests and procedures are conducted without solvents or process fluids yet introduced to the plant. For example, control system sequence tests.

- Cold Commissioning of plant and machinery: Cold commissioning is undertaken when the plant and machinery have been duly made ready to start actual operation, as intended. Purpose of cold commissioning is to make sure that no surprises will spring up upon putting plant and machinery to actual operation that they are intended to perform. To achieve this, the operational tests of the facility during cold commissioning are carried out with inert material or water.

- Hot Commissioning of plant and machinery: Hot Commissioning can begin after cold commissioning has been successfully completed. Hot commissioning involves handling the actual process fluids rather than inert fluids or water. Hot Commissioning is often also referred to as Wet Commissioning.

Obviously, Hot Commissioning needs to be considered at par with commercial operation of the facility, so far as safety and access controls are concerned.

All the commissioning process described above should obviously also have included the pollution control and effluent testing and monitoring system and it should have been established that all emissions are well within the permitted limits.

- Live Commissioning, Start-Up and Initial Operation: Once various individual islands of process systems have been tested with process fluids, it means that all the systems and plants of the facility are ready to undergo actual production as intended.

When this stage is reached often it may become necessary to bring into picture certain external agencies. For example, in case of a power project it may become to bring into picture the utility agency or the power grid control as well as the party or parties that will consume the electrical power to be generated. In case of a chemical industry the party supplying the raw materials will generally have to be informed through project owner so as to be ready for meeting demand of the materials during live commissioning, start-up and initial operation; and probably the party, supposed to be consumer of the produce, too may have to be brought into picture through owner of project.

In case of an electrical power project the electrical grid authorities would need to be brought into picture through owner of project since starting of motors of large capacity may bring about some current surge and also synchronizing of the incoming generator under commissioning may cause electrical surges, more so if during testing the generator is tripped from full load for checking efficacy of turbine governor controls by way of preventing occurrence of overspeeding of the turbo-generator.

Pollution Control Authorities must in any case be brought into picture, through project owner, well in time. The authorities should be satisfied regarding total compliance by the facility of the limits prescribed in the permits issued and this fact should be documented properly – preferably duly witnessed by pollution control authorities.

- **Performance Testing or Performance Guarantee (PG) Testing**: All projects are meant to achieve a certain minimum 'Performance' which is more often than not, well defined in contract. 'Performance' is supposed to be guaranteed, generally under 'penalty' (defined in contract) jointly or severally by an OEM and an EPC Contractor. In cases where there is no involvement of an EPC Contractor and actual execution is handled in-house through several separate entities or contractors, each with its limited scope generally 'Performance' pertains to each individual equipment or machinery, there generally being no concept of overall 'Performance'. In case of EPC, however, more often than not the term 'Performance' covers a lot more than performance of any single equipment or machinery. Therefore, requirements of 'Performance Testing' would be different for different types or manners in which projects are executed. Described below is a typical description of it (performance testing) for projects executed on EPC basis.

The figure of guaranteed performance, in all EPC contracts, is always a slightly, but significantly, lower than best possible. Project manager should so manage that while carrying out its own operations of facility for the purpose of commissioning makes sure that actual best performance of facility if significantly better than what is required under contract performance guarantee. Reason for suggesting this is that it even with best efforts such better than guaranteed performance cannot be achieved and the best that can be achieved at all costs is what is guaranteed in contract then there is every likelihood that some slight wear and tear (all equipment and machinery are prone to start degenerating immediately upon being started up and put into operation) may take place and affect actual performance such that the performance guaranteed on contract is not achievable. If this happens, effects can be very undesirable and detrimental for EPC contractor, who may end up paying heavy penalties that may not only reduce profit margin but also prove to be cause of occurrence of financial loss rather than profit coupled with loss of reputation that may affect future business.

After all plant and machinery of project has been duly fully commissioned and put to all operational checks in accordance with the applicable codes and project manager has satisfied himself/herself about the aspect of guaranteed performance being easily and certainly achievable, it is time for project manager to demonstrate to the owner that the plant or facility concerned is ready for performing as intended. For this purpose, project manager should give a written official intimation of his/her readiness to carry out performance test for being witnessed by owner.

For the purpose of physically carrying out performance test, proper protocols should be drawn in advance, project manager and owner being in full agreement about conditions of operation of the facility under which the performance test would be carried out. Also, important at this juncture is full clarity of method of computation of 'Performance' and the formula to be used for this purpose. Almost in all cases full clarity already exists in EPC Contracts and none of the parties is in any doubt about this aspect. However, nothing should be taken for granted and in case project manager is not able to lay hands on the method of computation of 'Performance' and the formula to be used for the purpose, it is highly advisable that this aspect is discussed by project manager, threadbare, with owner and recorded in protocol.

The protocol should also include details of instruments to be used for measuring various parameters necessary for computing 'performance' of the facility, plurality of sets of performance readings. Copies of calibration certificates of all instruments to be used for performance test should also be furnished to owner, under records.

Another very important aspect that the protocol must clearly include is what is required of the party to which project manager belongs, after the performance tests have been completed. This is because at the instant of completion of performance test expenses are being incurred in terms of input raw materials being consumed, manpower being deployed and production of the commodity that the facility is supposed to produce is occurring. Just after the moment of completion of recording of readings of various instruments deployed for carrying out the performance test, there is no longer any need for continuing to run or operate the facility, so far as the party to which project manager belongs, is concerned. Moreover, the instruments, specially installed for recording values of various parameters that are to be recorded for the purpose of performance test, are all high precision instruments not meant for being left in the installation for any further purpose and need to be safely removed. Therefore, it is necessary to clearly spell out, in the protocol, the activity that will follow completion of performance test. Furthermore, it is customary to safely shut down the facility on completion of performance test, unless indicated otherwise in contract.

Modern practice is to install intelligent digital instruments specially meant of the purpose of execution of performance tests. These instruments are hooked to a computer through a router. The computer has a Performance Test Program pre-loaded often with a provision of either giving a command via a push-button for recording readings or values as measured by each instrument at the instant of pushing of the button or alternatively even this function is programmed to be executed automatically.

Current practice in the year 2018 is to also install indicating type of instruments so that their readings are taken physically and noted down on paper for the purpose of cross checking of performance results by carrying out manual calculations. Therefore, it becomes necessary to deploy several persons, each located near one of the instruments for physically recording its reading, instant for simultaneously reading each instrument is announced either by a clearly audible beep or bell or through strategically located loud speakers, it being the least preferred method.

It is advisable that project manager welcomes as many as possible of owner's personnel for purpose of such manual recording of instruments readings as doing so often has better psychological effect of satisfaction of owner.

## Activities post Commissioning and completion of Performance Test

Reader is requested to recall that it was emphasized at the outset that foundation of all projects is the contract for their execution. This fact assumes prime importance at the stage when performance test has been completed. A project manager would automatically be able to make out what his next should really be.

There are always very high chances that the facility shall be required to be safely shut down upon completion performance test. However, project manager should not hastily take a decision to remove the high precision specialty instruments installed for carrying out of performance test. This is because just carrying out performance test does not suffice so far as conditions laid down in contracts are concerned because it is equally important for project manager to officially submit performance test results to owner and obtain their acceptance of the results. If, by any chance, there arises a dispute as to performance test results being acceptable, the performance test would have to be repeated and in that case if the high precision specialty instruments installed for carrying out of performance test have been removed, they would have to be reinstalled and hooked to computer. Ideally, therefore, it is desirable to leave the entire arrangement in place, unless doing so is totally unfeasible for any reasons.

Upon completion of performance test and official acceptance of test results, first priority of project manager should be to arrange to safely remove the high precision specialty instruments installed for carrying out of performance test and to restore the facility to its normal form.

## Management of Handing Over of project to owner.

The manner of handing over of project to owner is defined clearly in execution contract and project manager would by himself/ herself be able to fully make out the procedure to be followed.

Described in the following are some possible scenarios and corresponding requirements to be fulfilled. However, the following few tips may be helpful if contract provides for handing over of project to owner, in a ready to operate condition.

## Handing over project to owner in Ready to Operate condition.

The term "ready to operate" obviously means that project manager would, on completion of performance test, so arrange that lubricants and specialty materials needed for first fill are already filled in various systems of facility as needed. However, there are circumstances that are rather tricky. For example, in case of a large capacity modern thermal power plant the dry high purity Hydrogen Gas filled under pressure in generator cannot be left filled unless Hydrogen Seal System is operating which means that pumps pertaining to lubricating and seal oil system should be in operation and the generator must not be stand still – rather its rotor should be under continuous slow speed rotation by means of the rotor turning mechanism, which in turn means that operating manpower is deployed.

If this is a contractual requirement, it is rather tricky to handle. And project manager should, in such a case, confer with owner's representative so that owner's operating manpower is deployed, in all working shifts, side by side with project manager's manpower (i.e., contractor's manpower) so that as soon as handing over of project takes place (generally this happens in some office and pleasantries are shared with claps and photo shooting) project manager along with his/her operating team physically handover facility to owner's manpower with shake hand, smiles and thanksgiving. Scenario may not necessarily always be exactly as described here and project manager may have to handle the situation on a case-by-case basis.

## *Handing over project to owner in totally stand still condition.*

If contract calls for handing over of project in a totally stand still condition then the matters become even more tricky. This is because a plant that was operated, obviously at full capacity (or full load in case of a power plant) and under such circumstances, just putting it into a condition of safe total shut may technically not be right unless the duration, for which it is likely to remain in such condition of safe total shutdown, is clearly known. The reason for this is that just cooling down and stopping a plant or facility is likely to cause setting in of deterioration of machinery. Parts would soon start rusting, getting chemically corroded and degenerated unless the entire facility has been brought in a condition of scientific preservation. Machinery having rotating components that were subjected to very high temperature, particularly if such components are quite long and supported by bearings which are located far from each other – for example gas turbine or steam turbine, are likely to suffer bending and would like to given rotation from time to time or through a continuous slow rotation, as long as such machinery is not put into operating condition. Under such conditions technical prudence calls for putting various machinery and equipment in a state of preservation in a scientific manner and as recommended by respective OEM.

However, scientific preservation of any industrial facility is not only quite unlike mummification because scientific preservation of facility, if required under contract conditions, calls for deployment of some rudimentary bare minimum skilled and semi-skilled manpower, although only from time to time for inspections, recording of relevant details and often replenishing preservatives and cleaning the facility. Obviously, there would be substantial cost of implementing all this. Therefore, in cases where contractual requirements do call for putting facility under preservation, the cost aspects too are fully clear. However, if owner demands putting facility under preservation without it being a contractual requirement, it would be prudent for project manager to bring in picture the cost aspect of meeting such requirement of owner and properly settle the issue in a prudent manner.

If requirement of carrying out preservation of facility was provided for in contract calling for the preservation for long enough or an indefinite period, it would be prudent to sublet the job of preservation to a competent contractor, specializing in preservation of industrial facilities. If, however, requirement of preservation is only for a short enough period subletting thereof may not be required.

In all the above cases there, almost certainly, is a requirement of handing over to owner of documentation along with handing over of project. Importance of documentation has all through been emphasized in this book and it is expected that a project manager who has adopted systematized working must already have properly and systematically put in place all the documentation to be handed over to owner. It being so just a couple of documents may have to be added to the lot, at this juncture – mainly the details of performance test, computation details along with results and copy of acceptance thereof by owner.

## Spare parts & tools

Projects being executed on an EPC basis almost always also include supply of 'spare parts' in scope of EPC contractor. Project owners generally do not like to take over spare parts before taking over project because of two reasons; firstly, such taking over of spare parts would involve payment of agreed total price of spare parts for no productive purpose and secondly because owner would have to keep safe the spare parts and make facility spare parts store operational in all respects. Therefore, spare parts too have to be handed over to owner along with project. For this purpose, it is considered prudent for project manager to have a detailed list of all spare parts ready and handing over of spare parts should be carried out such that an authorized representative of owner does stock taking of spare parts after due inspection so that there may not remain any doubt about quality and acceptability of any of the spare parts.

Post handing over by EPC project manager sole responsibility of safe keeping of spare parts rests with owner.

Often contracts require certain consumables to be handed over. It is for project manager to have made sure that all such items are already kept in readiness, for being handed over to owner.

Certain special tools and tackles are used during installation of various machinery and plants. In general contracts specifically require their being handed over to owner. It being the case, project manager should arrange to put such special tools and tackles in clean working condition and with rust protection coating, in their respective carrying cases or alternatively a wooden box or two specially constructed for this purpose. If for any reason any of the special tools and tackles had got damaged it would be prudent for project manager to obtain a replacement for handing over to owner.

## Post-Performance Test handling of facility under BOO or BOOM or BOOT agreement.

BOO or BOOM or BOOT agreement require facility being commercially operated and maintained for sufficiently long period and involve complex commercial as well as administrative and legal requirements. It is but obvious that these aspects would have been examined under microscope before contract acceptance. There can be almost innumerous permutations and combinations of such commercial, administrative and legal aspects and discussion about them does not come under purview of this book as it pertains only to the topic of execution of projects. Therefore I would rather steer clear from complexities of commercial, legal and administrative dealings as a separate voluminous book can be authored to cover those topics.

So far as project manager is concerned, he/she should be able to find all relevant details of the manner in which a BOO or BOOM or BOOT project is meant, in agreement, for being operated and should take actions accordingly so that the Operation and Maintenance (O&M) Team can take over there onwards.

## Responsibilities/Services Post Handing Over of Project

The EPC Contractor or various equipment and machinery OEMs are required to take care of the following responsibilities and/or services. However, since these services pertain to long term dealing between EPC Contractor or various equipment and machinery OEMs, in general project manager who had been instrumental in execution of project is almost never in picture about these responsibilities and/or services. However, since the topic pertains to industrial projects, a brief description of certain basic requirements is being given here.

It is expected that project manager would be able to get an idea of the documentation that if not created properly at the stage of project execution and completed up to the stage of handing over of project, would never be available and lack of availability of such documentation may result in EPC Contractor or various equipment and machinery OEMs, as the case may be, having to bear costs towards responsibilities for either compensating project owner towards certain deficiencies or supplies and services that truly speaking would not have been included in scope.

## Warrantees during warranty period, as defined in contract.

Warrantee means a written guarantee, issued to the purchaser of an article by its manufacturer (OEM), promising to repair or replace it if it becomes necessary within a specified period of time (generally one year from date of manufacture; but often purchaser bargains for warranty period starting from the time of handing over of project), provided the article concerned is put to use strictly in a manner as defined in OEM's manuals. Often OEMs like to post a representative at project site after its handing over, up to end of warranty period. With the beginning of the Warranty Period (Defect Liability Period) the attention to close the remaining listed open points may decrease for some contractors. Therefore, a clear system of listing, reporting and settling of open points, either the ones attached to the Take Over Certificate or those arisen during warranty period, needs to be implemented and monitored on a regular basis by project management. Also, during defects liability period some changes on systems might be implemented, which then are forgotten to be documented in the as built documentation.

Some owners insist on inclusion of warranty from OEM for Latent Defects in equipment and machinery. Latent defects are those defects, which are undetectable during inspections and tests that are carried out during manufacturing process by way of quality control. Quite obviously, even OEMs are clueless about them. Normally latent defects remain dormant all through lifetime of machinery. Therefore, normally it is economical leave latent defects to be taken care of by insurance of facility during period of its operation, however without mentioning the words "Latent Defects" because even insurance companies hate this expression.

In case of EPC contract many times civil works too are covered under warranty, the warranty period generally being long (typically 5 years). There is nothing that project manager or EPC contractor can do to prevent any defect of civil works showing up during civil works warranty period. Such matters are dealt with on a case-by-case basis. One should note that for interest of the owner, the defect liability period should be systematically used to retrieve field data, which should be available to its end. Huge amount of data regarding failures, corrective maintenance works, turnover spare parts, design changes, etc. should be available for future reference.

It must be noted and clearly understood that no warranty comes without any disclaimer and owner cannot claim any benefits in an unjustified manner.

## Long Term Service Agreement (LTSA)

The LTSA by EPC Contractor or various equipment and machinery OEMs (mainly gas turbine manufacturer) can also be one of the requirements of contract.

LTSA requirements are invariably a part of contract in case of projects comprising gas turbine, such as power projects and some industries using gas turbines to drive compressors. This is because during operation gas turbines, being internal combustion machines, handle the thermodynamic fluid (truly speaking – flame which is combustion product) at exceptionally high temperature. These temperatures are such that all known metals would not only lose their strength but also melt. That gas turbines are able to operate successfully is due to ingenious methods adopted by various manufacturers for (a) prevention of direct contact between flaming hot combustion product and gas turbine internal components and (b) arranging flow of air (air is a thermal insulator) over and close to surfaces of components directly exposed to combustion product – each manufacturer adopting a different technique. Owing to this gas turbine internals require frequent inspection of vulnerable internal components.

OEM's experts, who thoroughly know what to look for in order to detect setting in of any abnormality that may result in failure during operation, perform such inspections strictly after specified durations and operating conditions. Obviously, such expertise is not available to any ordinary engineers and technicians belonging to normal machinery operation and maintenance teams and involvement of OEM is a must. If inspections reveal any developments of concerns (for example start of a crack) it becomes necessary to replace such components with refurbished or new ones as the case may be.

Further, in case of gas turbines there is a life cycle after which certain critical components need to be replaced with new ones and this activity is termed Major Overhaul. Generally, incidence of Major Overhaul takes place after about 15 to 20 years of commissioning if machine is used regularly depending upon whether operation all through has been smooth (with least plurality of shutdowns and start-ups) as well as least plurality of incidents of sudden heavy changes in load. Therefore, some owners like to deal with the incident of necessity of Major Overhaul when it comes to while others like to bind the OEM by including a Major Overhaul in scope of Gas Turbine OEM along with LTSA. Terms of LTSA agreement also clearly define responsibilities of project owner towards ensuring that only competent persons handle equipment, for operation as well as regular maintenance, and all actions and observations are well documented and archived for reference in the long run. At present computerization for operation of equipment has advanced in technology that so far and operation of equipment is concerned all necessary records get automatically by way of highly efficient acquisition of relevant data. However, records of various actions towards maintenance have to be made manually. Owner has to ensure that such records are created meticulously and thoroughly and are properly archived, so as to be referred at some time in future in the long run.

Discussing this topic any more would not really serve any major purpose for project manager who would in any case not be made responsible for handling LTSA on behalf of OEM. Therefore, we put the topic to rest here.

## Closing of project execution contract

Closing of project execution contract signifies the end of responsibilities of the party whose responsibility it was to execute contract, therefore, amicably closing contract is a matter of foremost importance for project manager. Opportune time for closing of project execution contract comes immediately after the date when last of all warrantees, under project execution contract, is over.

At the time of closing of contract certain final payments are to be released by project owner to the party responsible for project execution, in terms of contract for execution. Manner in which it is to be done is supposed to be defined in contract and it may become necessary for project manager to submit certain documents to project owner as a proof of project execution contractor having fulfilled its obligations under contract. It is for project manager to make sure that he/she takes all appropriate actions leading to closing of project execution contract.

Other contracts such as LTSA, which always is a separate Agreement (or Contract) should, in general, have no linkage with closing of project execution contract and therefore should be of no direct concern for project manager.

-: The End :-

# APPENDIX 1: CULTURE EFFECT

*"The stranger sees only what he knows"*
*(African Proverb)*

The scope of this Appendix is not to disclose in detail particular cultural aspects and experiences of working on global projects, in particular in developing countries, but to highlight strategies that project managers should take into consideration while managing projects when applicable. Possible culture shock feelings may be triggered in foreign workers while working in such alien environments due to the abrupt loss of cues and norms which usually oriented them in their daily lives. Members of different organizational cultures may have different norms, values and policies that may lead to misunderstandings, hidden agendas, which may bring lack of trust, uncertainties and even conflicts. It demands huge attention from project managers to respect and to understand these diversities, keeping and increasing the sense of participation of all workers without prejudices and lack of motivation. The unconscious presumption that there is one normal, single way of doing things or of socially behaving is a dangerous pitfall for inexperienced incomers in multi-national environments. Project managers should, however, not allow this to happen by means of stimulating proper integration among team members and setting the project culture values. It is to be highlighted that each culture has its proud heritage and therefore all respect is to be shown.

Besides the unawareness of cultural values of project team and stakeholder members, more aspects of contracting processes are at risk for cost and schedule overruns when different cultures are working together. Knowing how to navigate these constraints is a very important element in being successful in developing countries. Knowing how to fit in also requires knowing how to do so without breaking the rules and regulations of the home base country. Fortunately, most of these issues can be managed, and the risk of them derailing the project can be minimized accordingly. The issues include the following:

- Material and equipment deliveries can become unpredictable when passing through customs, as well as local equipment rental contracts;
- Foreign labor sourcing, in case needed, can become embroiled in issues related with availability of qualified workers and immigration rules (Visas and work permits), if local authorities create barriers during the contracting process of foreign workers, in order to encourage the use of the local resources;
- Quality and safety practices done in different acceptable standards
- Local political interference, which may include corruption practices for expediting paperwork, awarding contracts, and other issues that most contractors' home countries prohibit this.

Cultural differences may affect also the approach to safety and quality matters. In many countries safety is not perceived in the same way as it is in developed countries. The monetary impact caused by safety related incidents is not severe and the risks assessment more tolerant. Quality control also may be viewed as an impediment to the schedule and therefore relegated to the lower end of the project's priorities. The single driving factor is to complete the job per schedule and let 'tomorrow take care of itself'. The initial response to the disruptive effect that an injury or death can have on a project in some countries is that there are plenty of workers just waiting for a position to become available. Often, contractors attempt to provide safety training using their tried and true tools brought from home, tools such as the widely used training films from several companies with renowned safety records. Most of this is not absorbed by the local workers, even when translated into their local language. More effective are homemade training films using local workers as actors along with tools and equipment they will actually be using on the jobs.

Quality is also a thorny problem. Although the owner of the project will expect the contractor to do a quality job, the perception of quality may be different, depending on who is viewing it, even considering clear contractual quality standards. The owner may expect a certain level that exceeds the contractual and local norm. The contractor in trying to enforce these quality requirements may have difficulty in getting the local workforce to understand the requirements and the need for doing things differently. Often the attitude among local workers in that quality is just an impediment, schedule is the only requirement. With international code requirements not fully adopted by many localities, it is sometimes difficult for a contractor to remain competitive when the local standards are not up to usual standard he is accustomed to working with. Again, in this subject, the proper training of workforce in areas needed to catch up expected standards are required, and it is the project manager's responsibility to identify in due time such shortcomings and react properly.

Another aspect worth to mention, regards the cultural motivation effect that directly affects productivity on global projects. While working on global projects it is common that project managers have to manage workers from several nations throughout the world with different cultural backgrounds. Even within one country, several languages may be spoken, like in India, and citizens may practice different religions. Having a basic understanding of the culture of a host country and acquiring basic knowledge about local religious beliefs and habits are essential when designing methods to improve productivity on construction projects.

Some examples below, out of observations taken along several years acting in the international arena, show types of cultural issues that should be taken into consideration to avoid conflicts and to better adapt to foreign cultures:

- Countries like Japan, China and Korea, by the fact that they were invaded and controlled by another government in the past, due to their unique heritages, customs, languages and attitudes, should not be inaccurately generalized as the same. This mistake is misleading and causes conflicts;
- In every culture in the world, people do not like to be publicly ridiculed, and in Asian cultures, public ridicule is called losing face. If someone loses face at work, he might not return to work and if the embarrassment is severe people may commit suicide for the embarrassment caused to their firm, family and to their country. In some countries, it is rude to interrupt people while they are talking (Asia), but in other cultures if no one interrupts a speaker will continue until the meeting is over. Also, insults in Arab culture could be interpreted as an insult to an entire family, triggering revenges even very late.
- People whose native language is not being spoken

need time to translate, time to think about the answer and time to translate to their language. Tolerance is required.
- When one gives a presentation, the audience is judging him on their technical expertise and in relation to local standards. In most western countries professionals use charts, graphs and flipcharts and they write on them during the presentations. In Eastern cultures this is a sign that the presenter is not prepared for the presentation. Well-rehearsed should be the only type of presentation in Eastern cultures. In Asian cultures, a presenter should never sit on the edge of a table because audience could see it as a sign of disrespect.
- Meetings are conducted differently in every country, varying from autocratic lectures to free-form discussions. It is advisable to observe local meetings before, also to know if interrupting someone is acceptable or to wait the speaker to finish first.
- There are some gestures and behaviors that are unacceptable in certain regions of the world. Never touch, or give /receive business cards, anyone who is from an Islamic culture with the left hand. Women should never be approached or touched in Islamic cultures, because it could result in severe punishment to the women. Shouting and vehement refusals are part of Islamic business culture, and they are not signals that a business meeting is over, merely that more bargaining needs to take place. Giving in too early in a negotiation could be interpreted as a sign of weakness.
- In Western cultures people point with the index finger, but in Eastern and Islamic cultures pointing the index finger is rude. Even pointing with a foot (when legs are crossed) is rude, and tables have cloths around the front of them to hide the feet of people

sitting there.
- Hugging business associates is acceptable behaviour in countries in the Middle East, South America and France, but people from other countries just shake hands in business events. The length of time someone holds onto someone's hand also varies throughout the world. People from Arab cultures might hold the hand long and pump the hand repeatedly. Asian handshakes are gentle instead.
- People from Asian cultures may not use handkerchiefs in front of other people and sometimes they blow their noses onto the ground as they consider handkerchiefs to be disgusting.
- When working in an Islamic country, it is wise to hire a local driver. If a foreigner accidentally kills someone in a car accident, the penalties are severe, and it is best to leave the area immediately (sometimes leaving the country is advisable).
- In Islamic cultures, for example, people accept their position in life, and they do not live to work. instead they work to live. Since their status in society is determined by birth, their clan, their tribal affiliation, or their place to birth, work is not a means for elevating one's status, differently as in societies based on protestant work ethic, where people are taught that hard work will be rewarded by raises and promotions.
- For South America and Asian countries in general, cultures are paternalistic societies, where companies are like families and supervisors behave like parents, admired at work and treated with reverence by employees and workers.
- In countries ruled by dictators or benign dictators, people realize they have little power, so they may resort to passive-aggressive behaviour when they receive a bad evaluation or they are feeling out of con-

trol, or even due to large disparities in compensation between managers and workers. This behaviour is manifested by means of workers slowing down their work, missing deadlines, pretending not to know how to do something, becoming argumentative, pretending not to understand instructions and doing work incorrectly on purpose. Knowing the source of passive-aggressive behaviour helps a manager to figure out how to deal best with it.

- In countries which were colonized by European countries, foreigners still could be referred to as Sir (even if person is female). The title of Sir is used for all levels of foreign personnel, so it is not an indication of who is in charge.

# APPENDIX 2: TYPES & TENDENCIES

*The Indian system of Gunas*

To the extent that a much of project management in countries like India entails people management, it is helpful to be aware that humans have certain basic traits & tendencies.

According to an ancient scripture of India (namely, Geeta, Chapter-4, Verse No. 13) there are three basic tendencies present in differing proportions in each human. Based on these tendencies humans can be classified into four types.

For ease of understanding I would like to classify the types of humans as, Alpha (α), Beta (β), Gamma (γ) and Delta (δ). The three basic tendencies (called 'gunas' in the scriptures) that are present in all humans are:
- Sattva (Sanskrit language word for Divinity or Intellectual – is the quality of balance, harmony, goodness, purity, universalizing, holistic, constructive, creative, building, positive attitude, luminous, serenity, peacefulness, virtuousness. Sattva results in excellent health – both mental and physical and mental health slightly

dominating over physical, awakening, awareness, alertness, sharpness of brain, power to focus thoughts, power to think, power to co-relate things, memory retention and recall etc., under this state of mind other two tendencies (or gunas) namely the Rajas and Tamasa remain dormant in different proportions, Tamasa almost being absent). A person with dominance the tendency of Sattva is said to be Sattvika.
- Rajasa (Sanskrit language word for Ruling tendency or Managerial tendency – is the quality of passion, activity, neither good nor bad and sometimes either, self-centeredness, egoistic, individualizing, driven, moving, dynamic. Rajasa results in excellent health – both mental and physical but physical health rather dominating over mental, looking forward to what to perform and where, recalling and planning actions, strategizing actions, assigning tasks to others, inspecting progress etc., under this state of mind other two tendencies (or gunas) namely the Sattva and Tamasa remain dormant in different proportions, Tamasa being rather too feeble). A person with dominance the tendency of Rajasa is said to be Rajasika.
- Tamasa (Sanskrit language word for Sloth or Sluggish – Tamasa is the quality of imbalance, disorder, chaos, anxiety, impure, destructive, delusion, negative, dull or inactive, apathy, inertia or lethargy, violent, viciousness, ignorance. Tamasa results in somewhat weaker health – both physical and mental and physical health rather dominating over mental, tiredness, sleepiness, weak memory and weak mental retention, lack of inquisitiveness, lack of drive to perform and to compete, lack of will to excel in any activity or capability, under this state of mind the other two tendencies (of gunas) namely Rajasa and Sattva remain rather dormant – Sattva being most dormant and sometimes absent. A person with dominance the tendency of Tamasa said to

be Tamsika. Such a person may be found to be having a tendency, for example, of avoiding use of safety appliances meant for personal protection of persons under industrial environment. You may find a person who prefers to check whether an electrical wire is live or dead by touching its conductor with bare hand while moving the hand in a very swift manner, rather than using a Voltmeter or a Neon Indicator. Quite obviously once in a while some electricians of this kind die by trying this trick with high voltage cables.

No one is either purely Sattvika or purely Rajasika or purely Tamasika. Human nature and behavior is a complex interplay of all of these, with each tendency present in varying degrees and even the proportions of the three tendencies vary from time to time in the same person.

In some persons, the conduct is Sattvika with significant influence of Rajasika and Tamasika tendencies, Tamasika rather being dormant. In some persons, the conduct is Rajasika with significant influence of Sattvika and Tamasika tendencies, Tamasika being somewhat dormant.

Demographically the largest number of persons is Tamasika, the Rajasika persons being next in numbers and the Sattvika persons being the least in numbers.

Furthermore, these tendencies do change in every person during time of a day. Sattva or Divinity is most dominant early in the morning just upon waking up from a good and healthy sleep. That is why all healthy persons feel happy upon waking up, have a tendency to smile, look forward to think afresh and plan and progress.

Rajasa tendency tries to be dominant as the day progresses and late in the morning the Ruling or Managerial tendency tries to be dominant. That is why the period between early morning and midday is the one when the focus is strong on correct management of the affairs. This continues until lunch hours.

Tamasa tendency sets in after lunch (in some countries this tendency results in sleepiness making siesta a tradition, post lunch). However we humans have learnt how to overcome of significantly suppress the tendency of Tamasa by supplying the brain with stimulating chemicals such as caffeine (by consumption of coffee). Tamasa tendency, however, becomes most dominant as sun sets. This results in Night Life and the related activities.

From the above evolves a table showing relation between type of person and the jobs he/she can be expected to handle properly.

| Type of person | Main tendency | General tendency | Dormant tendency | Type of jobs for which such person best suited |
|---|---|---|---|---|
| Alpha (α) | *Sattvika* | Sattvika + Rajaiska | Tamasika | Research, Design and Engineering, Licking for exactness and tendency to express situations numerically, Planning, Scheduling, Visualizing Training Needs, Designing of Training Modules. |
| Beta (β) | *Rajasika + Sattvika* | Rajasika + Sattvika | Tamasika | Implementation of Design and Engineering, Implementation of Planning and Scheduling, Designing Monitoring and Reporting Systems, Management of Cash Flow, Planning of Finances, Administration. Training people, Visualizing special challenges with regards to safety needs for the project in hand. Handling jobs requiring very high skills. If engaged in Jobs needing high levels of technical depth, sometimes a Type Beta (β) person may apply theory somewhat mistakenly due to lack of full depth of technology. |
| Gamma (γ) | Rajasika + Tamasika | Rajasika + Tamasika | Saativika | Application of Design and Engineering, Application of Planning and Scheduling, Monitoring and Reporting, Assisting the administrator, Handling skilled jobs. Training semi-skilled and unskilled people, ensuring safety and implementation of safety. However, it is better to keep a Type Gamma (γ) person from going stray, through structured work charts that do not leave much to assumption and imagination. |
| Delta (δ) | Tamasika + Rajasika | Tamasika | Saativika | Carrying out semi-skilled and unskilled jobs, needing supervision from time to time. |

| | | | | | Type Delta (δ) persons tend to treat safety harness as a burden and despite having been provided with good quality safety shoes, tend to move about barefooted thus inviting hurt and even loss of man-days or in worst case lost of body parts resulting in trouble for project manager due to legal complications and expenses involved in providing the needed medical treatment. |
|---|---|---|---|---|---|

# APPENDIX 3: MANAGING HEALTH AND SAFETY PERFORMANCE

*"Safety first. Business next" (Unknown)*

Although ensuring that good safety practices is first and foremost a moral obligation for any corporation anywhere in the world, as it was disclosed in the previous appendix and along the text, this sometimes is cultural perceived differently in some developing countries. One should consider that besides this moral obligation, a lot of visible and invisible costs are associated with unsafe environments, which may badly affect project profitability. These costs go from clear medical treatment expenses; increased insurance premiums; as well as hidden costs (lost time of injured worker; inefficiencies due to temporary decrease in the moral of total jobsite; damage of tools, equipment and material installed; etc.). These hidden costs of accidents is not so easy to calculate and therefore is frequently overlooked.

Based on the above, it is project manager's responsibility to somehow implement the so-called Safety Culture within its jobsite, and disseminate this culture towards the whole team members, including level tier of contractors. A positive Safety Culture means an environment that safety plays a very important role and is a core value for those who work for the project. This means that everyone within the project feels responsible for safety and pursues it on a daily basis; workers go beyond the "call of duty" to identify unsafe conditions and behaviours, and are comfortable intervening to correct them, no matter the hierarchical position he occupies. This type of behaviour would not be viewed as over-zealous but would be valued by the organization and rewarded. Creating a strong and sustainable Safety Culture takes time and effort.

Communication is the primary and most important tool available for keeping it always reminded, via constant training; updated safety statistics; root cause analysis, in case any safety recordable incident happens; safety initiatives and programs (e.g. Safety day; Zero Harm); safety incentives & bonuses; safety procedures and risk analysis (MSRA); etc. All these management tools are good practices observed in successful projects. However, the most important ingredient, without which any safety oriented initiative would be worthless, is the clear and visible commitment from the whole management structure in this crusade, providing an example in all hierarchical levels that everyone working on the site must abide by. A statement like ´´Prevention of injuries or illness will be given precedence over operating productivity´´, coming from high hierarchical management levels, is a very good start.

Planning safety into a project is just as important as setting production schedules and planning for the delivery of equipment and materials. There is no substitute for thorough pre-task safety planning. Each step of each contractor's proposed work plan should have a safety element that address anticipated hazards and how to eliminate or guard against them. Also, as part of pre-job activities, safety training should be established. Most projects will require an initial site-specific orientation for each and every worker that comes on site, as well as for anyone, regardless hierarchical level, even visitors, to be allowed to be on-site. Further trainings are also to be organized as precondition before starting activities with specific situations, like: confined space access; forklift truck operation; hazardous material handling; permit to work; working at height; etc.

Some other aspect which have direct implications with safety at construction sites, and sometimes is overlooked by project managers, is the housekeeping. Site poor housekeeping is an issue that affects everyone, it is morally depressing and physically unsafe.

The most effect way to keep a site clean is for each employer to train workers in housekeeping and to emphasize it in safety meetings and during lunch breaks or shift changes. For example, a separate housekeeping crew, with workers from each major employer, is often used to make regularly scheduled rounds of the premises to police the area. The main contractor or owner, by its turn, provides containers at strategic places at site, where the collected trash is to be dumped by each of contractors, and is responsible to remove from site the accumulated trash when containers are full. Flammable, hazardous or dangerous materials will require respective areas to be barricaded. These waste material / trash storage areas often this is under supervision of the safety officers on-site, but it can also be led by a foreman from one of the contractors, rotating from contractor to contractor on a weekly basis. The price is small if compared to the cost of the inefficiencies and the impact from a safety issue.

One of the elusive goals of all construction safety managers is to score zero injuries at site. The implication is that by eliminating all at-risk-behaviour zero injuries will result. This Zero

Harms Program requires a change in culture and in thinking of management and employees. It requires a change in the belief that injuries are inevitable; it requires a belief that although injuries will occur, this does not mean they must occur! This concept says that setting goals of anything more than zero sends the message that some injuries are acceptable. Research has shown that there is always something else, or something more, that can be done to be safer. Practically the pre-condition for this program to be implemented requires the following:
- Risk Assessment of any site activity deemed relevant;
- Safety orientation and continuous training;
- Written safety policies;
- Written safety programs;
- Worker involvement, meaning adequate site leadership of project manager;
- Clear regulation for Rewards for outstanding behaviours and Punishment, in case deviations were practiced intentionally;
- Accurate safety statistics;
- Accidents root cause investigations, including near misses;
- Demonstration commitment to safety by top management

All site management, the supervisors and the foremen, should be aware about safety issues related with local worker practices that may affect global-construction projects. Some working practices, although all training given to workers, are deeply carved in daily behaviours and require some discipline measures to be re-educated and changed in daily activities. Such as:
- usage of regional forms of scaffolding, such as bamboo scaffolding that is tied together with nylon or jute rope, which is used throughout Asia;
- workers not wearing shoes because it is easier for them to climb bamboo scaffolding or workers are more com-

fortable going barefoot in hot climates;
- workers refusing to use PPEs, or even selling outside construction premises his personal protection devices;
- workers compensation frauds due to false claims in case an injury happens. Such fake claims could be associated to false injuries; aggravation of previous ailments; exaggeration of severity. These kinds of claims are very costly to the insurance and construction industries. Direct supervision of the claimants are to be vigilant about behaviors that suggest possibility of fraud, such as: short-term employment history: experiencing financial difficulties; early Monday or late Friday injuries; unwitnessed injuries; details of accident are vague; facing firing or lay-off. Good communication between supervisors and the worker, and between the worker and worker´s peers will usually put light on these issues. It is important that all alleged injuries be challenged. Since fraudulent claims are usually very costly claims, if they are not dealt with promptly and properly, they could go on for years;
- intentional injuries to opposing clan or tribe members;
- unsafe usual practices in lifting operations.

As final note, attention is required to prevent health problems in the construction industry. The most common hazards are related with heat, dust, aspiration of mineral particles or asbestos, noise, dermatitis, burns and toxic chemicals. All these hazards are preventable. Job sites in tropical areas may use mobile foggers to spray insecticide to kill mosquitos that may carry the malaria parasite. Pesticides that are banned because they are carcinogenic are still in use in developing countries. Occupational diseases are also a serious problem in construction, mainly in developing countries, with its direct costs and claims increase if measures are not taken to remove them, such as asbestos that is still legal in some developing countries..

# APPENDIX 4: MANAGING PROJECT RISKS

*"One of the many uses of risk analysis is in distinguishing between bad luck and bad management" (R. Raftery)*

Experience shows that it is not easy to successfully manage industrial projects in developing countries. Risks which have not been identified and managed are undoubtedly unchecked threats to a project's objectives, which in turn may lead to considerable overruns in cost and scheduling. A myriad of culprits could be evoked to explain this unfortunate fact. However, the most preferred one, used mostly by construction companies from developed world which operate overseas, is the famous "unpredictability" of such environments which forbids these projects to be successful, allegedly due to matters mostly related with local socio-cultural and economic / political aspects. Although such contextual aspects constitute areas of uncertainties, to which some important remedies are disclosed in this manuscript, it should be noted that for industrial projects the notorious delays, cost overruns and quality issues are mostly associated with the operational aspects related with the daily management, and therefore much more influenced by internal factors than external ones.

In other words, these operational caused uncertainties, and the due consequences thereof, such as unclear responsibilities, unrealistic planning, unclear procedures, inadequate tariff structures, inadequate design, inadequate staff qualifications, poor productivity, defective materials, etc., could therefore be reduced by means of better effective management implemented since the early stages of the projects, and by means of progressive monitoring based upon historical information and 'lessons learned'.

Therefore, the authors are in the opinion that disregarding the external risk factors that are completely out of control of the project managers, which for sure may have high magnitude of impact on industrial projects anywhere actually (e.g. volatile economic environment; inflation; price fluctuation; financial failures; political climate; natural catastrophes; etc.), the remaining internal risk factors are manageable, provided that the appropriate management effectiveness is in place. Also, the simple use of contingency sums to deal with risk is unlikely to encourage more effective management of projects, nor to lead to greater efficiency. Stakeholders (contractors, consultant, owners) need to have a more comprehensive understanding of the nature of the risk they face, their likelihood of occurrence and their potential impact on respective organizations.

Notwithstanding the appropriate due diligence performed in the early stages of the project, in order to avoid or mitigate impacts as consequence of the above mentioned risks, one should accept that no matter how thorough the planning, no matter how carefully and thoughtfully the plan was developed, certain adverse impacts of the unforeseen or under-predicted events may happen which should be accommodated in some fashion.

There may be strikes or inclement bad weather or bankruptcies not previously anticipated, which may involve additional costs, delays or scope creep, in case additional scope of work (or changes) is to be considered.

Therefore, contingency planning must form part of the overall planning process. This prudent management tool can take form of calculating a monetary cushion, usually adding arbitrary percentage to the bottom line, or even adding additional days or weeks to the schedule. But whatever form it takes, contingencies must be managed and it will constitute precious historical information for future similar projects.

The two most disruptive issues are usually project delays, caused by for instance late project start (late permit, problems with funding or bonding, etc.), and parallel projects, both draining the available planned skilled resources to other projects being executed at the same time. One typical example mostly used by foreign companies working in developing countries is the adjustment of labour productivity, by means of adding adequate labour hours on specific critical activities, considering the local environment or local expertise / available tools in executing such tasks.

In summary, managing the day-to-day industrial projects is complex in and of itself. Issues can arise in spite of the best efforts of the project manager. Issues also can arise due to direct actions or inactions along the project, causing unexpected changes. The tools to proper manage the potential changes are due diligence, claims and insurance process.

The claims process is very dependent on the contractual clauses agreed between parties on the early phases of the projects, which should become familiar to all stakeholders. As the project progresses, communications must be made an integral part of the process whenever an abnormal condition arises, forcing the parties to review the issue and look for the ways to overcome it. Allowing issues to move forward unsolved often results in costly claims and disgruntled people. Managing risks require intelligent planning, to consider open-mindly possible areas of concern, smart project managers, to understand and to process potential impacts, and fluid communication through project participants, to anticipate events and respective counter-measures. The unexpected can be tamed!

# APPENDIX 5: ENSURING QUALITY

*" A job worth doing is a job worth doing right" (Unknown)*

The construction industry in developing countries is often characterized with low productivity, lack of standards and poor quality. Unlike the manufacturing industry which operates under a closely monitored environment, where it is possible to control all variables that have bearing on product quality, construction on site is not an automated process and uses people rather than the machines and technology to influence the project outcome. The very nature of construction appears to be the real barrier to quality management success. Several studies on construction industry conclude that its fragmented nature, lack of coordination and communication between parties, adversarial contractual relationships, and lack of customer focus inhibit the industry's performance. Traditional quality management systems are often unrepresentative of workforce and, are usually preoccupied with instruments of control and its administration rather than the outputs that are important to the customers. The ISO based generic approaches to quality management are more bureau-

cratic in nature. These methods are generally abstract and more concern with the management system than with the control of the work process. Notwithstanding the foregoing, there are several factors that impede the management of construction quality in developing countries, such as:
- Lack focus on quality in contractual provisions, not putting emphasis on this criteria for pre-qualification of bidders, instead privileging lowest bid prices, unrealistic schedules, unachievable specifications and inadequate compensation to contractors against escalation.
- Organization structure of contractors with an ad hoc approach towards resource mobilization without no organization setup plan.
- Lack of technical capabilities (human and non-human) to ensure effective quality assurance on construction sites, as consequence of lack of financial capabilities which forbids adequate training of personnel and appropriate tools.
- Slow pace of mechanization in developing countries, with a labor intensive approach, rather than owning or renting expensive machineries, and as consequence lower productivity and quality.
- Lack of training and skills, relying on informal and traditional apprenticeship where labourers learn the trade skills from the master craftsmen. However, such training may not satisfy the demand for higher quality.

Taking the above into consideration, the same note regarding the success of safety culture implementation applies for the quality: all begins with the top corporate officials' commitment which flows throughout the entire organization. This commitment is visible by means of a policy, where the quality assurance is defined, and quality plans, where the rules of the quality control with respective quality standards are to be followed.

One should note that the setting up the quality organization requires some forethought. It should encompass also off-site activities, like the items supplied that shall be shipped to site. It is of the utmost importance especially when suppliers are not used to providing parts with required higher standard levels, or even longer lead times for delivery. In this case not only the shops have to be regularly visited during manufacturing process, but also its respective quality plans have to be vested to assure the required standards will be achieved. The cost to repair potential manufacturing defects on site can be tremendous in comparison to fix them while material is still at the shop. Another aspect to consider is that a sudden increase in defective work in any phase of the process requires a thorough investigation by the quality department in order to understand and if possible eliminate the root causes. This particular situation is usual to happen when refers to welding activities. It is responsibility of the quality group to maintain accurate statistical records of their NDT inspections. Similar to measuring the productivity for the purpose of increasing the output, measuring quality is for the purpose of reducing defects. Weld defect rate of welders will identify which welder is experiencing unacceptable rates, therefore a re-training should be considered, before further cost and time impacts happen.

Another very important element to ensure quality at work in accordance with all requirements are the audits. Generally, there are two types of audits: internal (or self-audits) and external (or third-party audits). Both types have the same goal: assessing whether quality controls are being followed. The internal audit is performed by personnel from within the organization being audited, but not by those who are performing the activities that are being audited. This internal audit plan should be devised taking into consideration the importance of the activities and areas to be audited, as well as the results of previous audits, and should be reviewed after each audit and updated if necessary. Its results should be documented and analysed for quality trends, and forwarded to senior management. The external audits should be used to remove any possibility of bias that may occur during internal audits. By the same token, a plan is to be developed with a third-party and the auditing team should be given the flexibility to go beyond the plan and investigate any other areas or processes that they believe may impact the work they are auditing.

As final words about this topic, I would like to emphasize that the sooner the defects are discovered, the less costly the non-conformance repair costs will be. This is particular valid at any phase of projects, but moreover when takes place transition from erection phase to commissioning phase. Since sometimes defective works may not be readily apparent upon turnover - usually referred to as latent defects - the cost for repairing this work, especially if it is not discovered until after the contractor leaves site, can be very high. Besides the obvious re-work costs for the repair, it may include also general delays on the job, even in turnover the plant with subsequent LDs penalties. Therefore, it is imperative that project managers understand their responsibilities when it comes to doing it right the first time.

In some ways, the engineering design process, including the pre-bidding phase, is the most important area where quality improvement systems can be applied. Various quality control initiatives performed on early phase of projects like ensuring a construction site has sufficient lay-down area; improving contractor/supplier qualifications; having accurate and transparent bill of quantities and better preparing bids; analyzing constructability aspects etc. contribute to enhance the quality during construction phase and to reduce non-conformance costs. The further down the development process an error goes before detection, the more expensive it becomes to fix. In many companies, the implementation of quality control in engineering design has lagged behind when compared to later phases of construction, an oversight that can cost businesses huge amounts money and time.

# APPENDIX 6: EFFECTIVE PROJECT CONTROL

> *"To improve productivity, you must manage; To manage effectively, you must control; To control consistently, you must measure; To measure validly, you must define; To define precisely, you must quantify" (Unknown)*

Construction projects continue to experience delays, low productivity, insufficient quality, poor coordination and costs overruns, regardless where the projects are located. Statistics say however it is more common in developing countries because existence of many problems that contribute to this phenomenon. The goal to be pursued for effective project control (time, costs, resources) is to have a tool that allows timely notification to project managers of impending issues before they become major in order to implement preemptive actions to re-orient trending job out of its collision course. It is not purpose of this manuscript to present project control methods. Instead, just an overview about the experienced and successful practices.

Several sophisticated Critical Path Method (CPM) programs are available for this purpose, and its successful application depends mostly on the accuracy of the data inserted and the ability of the practitioners to process them to assure good results. However, it is argued that the traditional CPM does not prevent wastes during the construction processes, even though attention was taken during planning phase to assure a full scaled Work Breakdown Structure (WBS), proper link dependencies, resource leveling and earned values project control techniques.

As 'waste' one should understand as construction inefficiencies in general like: quality defects, poor productivity; poor coordination; waiting times; over-processing activities; reworks; etc. The reason behind this evidence is that CPM does not track down lower levels of planned and executed activities, agreed on daily coordination meetings among site managers and supervisors, which define utilization or availability of resources (work crews, materials, equipment sharing, etc), site accesses, working areas, sequencing of works, or project general information in general, and allows early detection of impending disruptions of upcoming activities that may impact main CPM master schedule. Improper management and control of these potential site contingencies cause wastes which have to be minimized on daily basis decisions during site coordination meetings, therefore its management is of the utmost importance to keep projects under control, besides the traditional CPM tool. In fact, the CPM scheduling has a focus on strategic planning, with calculated floats, critical paths and main logic dependencies. The lower level daily planning, called lean construction (Last Planner System - LPS), deals with best usage of resources and site contingencies to minimize the wastes in order to best pursue the execution of the CPM plan. By the same token, it allows to increase the visibility of potential hindrances, and to propose timely remedies of the upcoming tasks, and all data will be fed back into the master CPM schedule in order to re-evaluate the actual critical path and to allow the implementation of timely preemptive actions, if needed. Both methods are complementary to each other.

Another aspect to be considered for the sake of project control is the continuous measurement of manpower productivity. The concept of labor productivity improvement has remained highly elusive since labor productivity rates are defined in a variety of different ways, ranging from the value of gross output per worker (labor hours or work hours) to careful attempts to measure the physical output of labor taking into account other factors that affect production. Here below, as an example, some categories of work that can be readily measured:

- Electrical Works:
    - Cable trays / Conduits installation: in meters / day
    - Wire, cable pulled: in meters /day
    - Cable glands and terminations: in units / day
    - Cable insulation tested: in units / day
- Mechanical Works:
    - Piping installation (large bore; small bore; carbon steel; stainless steel; alloyed steel): in meters / day
    - Piping welds: in Dia Inch / day
    - Piping Insulation: in linear meters / day
    - Ducting installation: in meters / day
    - Flat surfaces insulation: in square meters / day
- Civil Works:
    - Steel Structure installation: in tons / day
    - Steel Platforms installation: in tons / day
    - Excavations completed: in cubic meters / day
    - Piles driven: in units / day
    - Reinforcement installed: in tons / day
    - Concrete poured: in cubic meters / day
    - Cladding works (double skin; single skin): in square meters / day

For this sake, contractors shall submit actual quantity progress tabulations on weekly basis against the planned quantities in accordance with baseline schedule. Among these productivity ratios the most relevant to power plant construction is the welder's productivity, since welding works is often on the critical path of resource loaded schedules. Earning rules, with weights assigned to each activity step, determine how progress is awarded for the actual works performed and an overall efficiency factor per contractor can be determined.

Of course, there are several aspects that affect the labor productivity and it varies from country to country. These factors can be broadly related with management (proper planning; realistic scheduling; adequate coordination; suitable control); labor (union agreements; restrictive work practices; absenteeism; turnover; availability of labor and skilled workers); government (regulations; environmental rules; political ramifications) and local aspects (climate conditions; labor practices; acceptance of equipment). To improve productivity, the impact of each of these variables have to be taken into consideration for planning resources and management methods, and to select appropriate technology. Normally construction firms that are frequently involved in global contracting develop their own in-house base productivity factors for locations where they build projects.

# ABOUT THE AUTHORS

Mr Luiz Baptista and Mr. Anand Kumar Gupta met during execution of gas based power projects of Torrent Power Ltd., in Gujarat, India where the farmer was Project Manager representing the EPC Contractor and the later Executive Director representing the project owner or the customer. Both being of similar attitude with singular aim of making the projects grand success discussed and took liking of each other and resolved to work in unison with close coordination, which brought success.

When Mr. Gupta thought of authoring this book, he requested Mr. Luiz to co-author it so that the reader may be benefited to maximum extent. Mr. Luiz too wanted to share his experience with the needy and readily agreed to the request of Mr. Gupta.

The result is this book – aimed solely to orient those who may have to manage projects in developing countries such as India.

## Mr. Anand Kumar Gupta

Anand Kumar Gupta has long first-hand experience of over 51 years, at various levels, in industrial projects in India. His experience covers wide range of activities and aspects of projects covering conceptualization and development, specification writing, technology selection, land selection, international tendering, bid analysis, design & engineering, execution of project, testing & commissioning and profitable commercial operation post commissioning, based on project specific needs.

He graduated as a Bachelor of Electrical Engineering (with Telecom as an elective subject) from College of Engineering and Technology, AM University, Aligarh, India, in early 1963, a period when India had negligible industrial development with just a couple of small power plants supplying electricity in a few cities. Cities that had electricity even in British times mostly had dc power supply and those which were being added on the map of electrified cities were being provided ac electricity supply.

Telecom services were limited to Manual Telephone Exchanges in some cities and automatic dialing (Strowger Switching System) in the most important cities like Delhi, Bombay. Calcutta, Madras and some other developing cities. Work had just started on Microwave connectivity but in a small way.

There was just one National Radio Station named Aakashwani broadcasting on Short Wave Frequency with several branch Radio Stations in State Capitals broadcasting in Medium Wave Frequency.

Mr. Gupta utilized this situation to the best of his advantage and rather than limiting himself to electrical engineering & telecommunications, he indulged in practical as well as theoretical aspects of mechanical engineering, structural and civil engineering, keeping in touch with academicians as well as technicians.

The first job that he got, in 1963, was to install, commission and later on maintain Power Line Carrier-current Communication (PLCC) System of the Electrical Power Grid System pertaining to Rihand Hydro-electric Power Station of 300 MW in the state of Uttar Pradesh as also managing protective relaying of various distribution substations. Mr. Gupta rendered valuable services to the employer until the end of the year 1966 and switched over job to an upcoming captive thermal power station pertaining to India's first and now largest Aluminum Manufacturer (HINDALCO).

Mr. Gupta was taken by HINDALCO for PLCC System between their captive power plant and the aluminum factory and protective relaying of the upcoming power plant. However, post commissioning of the power plant, Mr. Gupta started taking interest in problem solving pertaining to all fields of technology involved in the thermal power plant and mastered upkeep, repairs and maintenance of coal handling plant, bi-cable aerial ropeway between mines & the power plant, coal pulverizers boiler tubes, heat exchangers, fans, pumps, valves, fly ash mechanical and electrostatic precipitators, steam turbines (suffering from blade failures in LP stage due to off-frequency operation) and many more.

As a result, Mr. Gupta was invited to join a well-known consulting engineers' firm based in New Delhi, India, as project coordinator, serving them for 2 years (1977 – 1979) whereafter he was invited back by Hindalco to manage project of expansion of their captive thermal power by adding two more generating units. During his second job at Hindalco he got multiple chances to travel to many countries of Europe, USA and South Africa in connection with inspections, technology selection and obtaining first-hand information about performance of products of various famous manufacturers. He utilized the experience of his observations from international tours for his project development work at Hindalco and succeeded in achieving glaring results. Post commissioning the thermal power plant project performed with over 96% Availability and over 96% Output Factor. Such exceptional performance brought Mr. Gupta in lime light.

In the year 1996, he was invited by famous Indian engineering conglomerate named Larsen & Toubro Ltd. (L&T) to head the power plant operations & maintenance sub-unit for their power plant EPC business. Mr. Gupta served L&T from 1996 to 2001, retiring from L&T in 2001 at age 61.

In the year 2003 Mr. Gupta was invited by Torrent Power Limited (TPL), a reputed power utility of Gujarat State in India. He was taken as Executive Director to manage their power projects comprising two large gas based power plants, to be executed on EPC basis. At TPL Mr. Gupta handled two projects of capacity 1500 MW and 1200 MW. He took retirement in the year 2014 at age of about 74 years.

Mr. Gupta has authored this book as a debutant with a view to share his vast experience of executing industrial projects in India, with a view to throw light on the path to be traversed in developing countries like India, where most number of industrial projects are likely to be executed in the near future.

## Mr. Luiz Baptista

L uiz Baptista (Mr. Baptista) has to his credit Dipl. Mechanical / Nuclear Engineer, with post-graduation in Project Management and Business Administration, as well as holder of NEBOSH and OSHAS HSE international certificates.

He is a highly experienced General Site Manager with demonstrated success of working globally in infrastructure projects (mainly power generation / transmission, railways and shipbuilding). He has been responsible for the execution of several projects under large EPC, TFA and Consortium contractual agreements, covering the bidding, the procurement, the execution and the commissioning project phases.

Mr. Baptista is skilled and focused in Planning & Control, Quality Assurance and Contract & Site Management, and with excellent HSE track record during project execution in different and challenging global locations. Due to his very rich and long work experience in the global arena, which includes several developing countries, among them Brazil, his own place of birth, his personal and professional background have substantially broadened the scope and target of this manuscript to other work environments, namely projects in developing countries in general, with similar challenges and pitfalls here presented and discussed.

Mr. Baptista throughout his career, which began in 1981 after his graduation, has worked in various positions related to the execution of projects, such as Site Mechanical Engineer, Welding Engineer, Quality Assurance Inspector, Scheduler, Planning and Control Manager, Contract Manager and General Site Manager. In 1992, he and his family moved from Brazil to Portugal, where they live, and when he has joined SIEMENS AG ever since, continuing his ongoing career.

www.ingramcontent.com/pod-product-compliance
Lightning Source LLC
Chambersburg PA
CBHW021812170526
45157CB00007B/2556